PROBLEMS OF ANIMAL BEHAVIOUR

By the same author

Animal Behaviour (Pitman Publishing Limited, 1985)
Editor of the *Oxford Companion to Animal Behaviour* (Oxford University Press, 1981)
Quantitative Ethology (Pitman Publishing Limited, 1981)

PROBLEMS OF ANIMAL BEHAVIOUR

David McFarland
Animal Behaviour Research Group,
University of Oxford

Longman Scientific & Technical
Copublished in the United States with
John Wiley & Sons, Inc., New York

Longman Scientific & Technical,
Longman Group UK Limited,
Longman House, Burnt Mill, Harlow,
Essex CM20 2JE, England
and Associated Companies throughout the world.

Copublished in the United States with
John Wiley & Sons, Inc., 605 Third Avenue, New York, NY 10158

First published 1989

British Library Cataloguing in Publication Data
McFarland, David, *1938–*
   Problems of animal behaviour.
   1. Animals. Behaviour
   I. Title
   591.51

ISBN 0-582-46820-5

Library of Congress Cataloging-in-Publication Data
McFarland, David.
     Problems of animal behaviour/David McFarland.
        p.   cm.
     Bibliography: p.
     1. Animal behavior.   I. Title.
   QL751.M394   1989
   599′.051 – dc19

ISBN 0-470-21209-8 (USA only)

Set in Linotron 202 10/12 pt Plantin

Produced by Longman Singapore Publishers (Pte) Ltd.
Printed in Singapore

Contents

CONTENTS

Preface

In this book I air a set of problems of animal behaviour which are basically motivational problems requiring some sort of functional explanation. The aim is to explore various aspects of the function-mechanism relationship. The problems remain, to my mind, largely unsolved. I have, in places, made suggestions as to possible solutions, in the hope that these may be taken up by someone with a fresh viewpoint, and fresh enthusiasm.

I hope that the book will be read by advanced students of animal behaviour, and that they will be stimulated to disagree, and investigate. Parts of the book have been read by some of my own students and colleagues, some of whom remain in disagreement with me on various points. The views expressed in this book are entirely my own, as are the errors. I thank all those who have read some of the early versions, for their comments, helpful and otherwise. I thank especially Charles Amlaner, Nigel Ball, Marian Dawkins and Robin McCleery.

David McFarland
January 1988

CHAPTER ONE
Motivational priorities

The survival and reproductive success of individual animals depends largely upon their use of resources, such as food, territory, mates, etc. At any particular time an animal may have alternative means of possible resource-utilizing activity. Every activity will have associated costs and benefits, in terms of the ultimate reproductive success, or fitness, of the animal. For example, a male stickleback, which moves away from its nest to court a female, may be increasing his chances of successfully luring the female to his nest and inducing her to deposit her eggs there, but he may be endangering his nest by neglecting to guard it against marauding neighbours, or to repair it after inadvertent disturbance by other animals (Wooton, 1976; McFarland, 1974). Thus there are both benefits and costs associated with the courting activity.

In general, the costs and benefits are attached to both the behaviour of the animal, and its internal state. Suppose we consider the situation of a deer, in cold, wet weather. The animal has a choice between standing up to feed and sitting down to shelter from the wind. There are costs and benefits directly associated with the behaviour. Thus by standing the animal has a good field of view and can easily look out for possible danger. It can feed, which it cannot do effectively while sitting, but it is much more exposed to the weather than it would be if it were sitting in a sheltered place. By sitting the animal is able to save energy, both by being less active, and by reducing heat loss in sheltering from the cold wind. On the other hand, it is less well placed to spot possible predators, or rivals, and it cannot easily feed. Clearly in assessing the relative merits of the two possible activities, the animal's state of hunger should be taken into account. If the animal is not very hungry, it can probably afford to sit down and wait for the bad weather to pass. If, on the other hand, it is in need of food, it may endanger its life by neglecting to feed now, since it may have to flee from a hunter, or other danger, at some time in the future, when it would otherwise have fed.

Even for such a simple decision as sitting down, or standing up, the balance of costs and benefits may be a delicate one. Account must be taken, not only of the costs and benefits of alternative behaviour patterns, but also of the internal state of the animal, and of the way in which the future behaviour is likely to affect that state. What an animal ought to do in a given situation (assuming that the overall objective is to maximize fitness) is to choose the option with the greatest net benefit (or least net cost). This kind of problem, generally called an optimality problem, has been the subject of considerable interest in recent years, and there is now a large literature on it (e.g. Maynard-Smith, 1978; McFarland and Houston, 1981; Stephens and Krebs, 1986). However, I am not primarily concerned here with what an animal ought to do, but rather how it does achieve the optimal outcome (if it does). I see this as an exercise in the theory of motivational mechanism, or proximal causation, whereas optimality theory is essentially an exercise in answering a functional question. However, I believe that the motivational question cannot be answered without reference to the functional question. In many areas of biology it is helpful to understand the function of the material being studied, but in decision-making it is essential, because decisions inevitably reflect a compromise among various different design features of the animal.

Conventional wisdom assumes that animals do that activity which they are most motivated to do at the time. Thus if an animal is more hungry than aggressive it will eat rather than ward off a rival. If a predator appears, fear will quickly dominate all other motivations, and the animal will flee or hide. This simple model of motivational competition raises an important and fundamental question: how are animals so organized that they are most motivated to do what they ought to do at a particular time?

At any particular time the necessary motivational conditions for many incompatible behaviour patterns will exist simultaneously. These conditions will generally include some aspects of the animal's internal state, and some aspects of its assessment of external stimuli. This interaction of external and internal factors has been the focus of considerable research, and has been formulated in various ways (see Toates, 1986, for a review). The consensus is that both internal and external factors contribute to the 'tendency' to perform a particular activity. It is also widely recognized that interactions among different motivational systems occur at many levels (McFarland, 1971, 1978; McFarland and Houston, 1981; Toates, 1986). However, there must always be a "behavioural final common path, which . . . involves the last type of interaction . . . in the causal chain, the final 'decision' before a potential activity becomes overt" (McFarland and Sibly, 1975).

2

Somehow, the animal has to make a 'decision' among the different possible courses of action. Moreover, such decisions must be made in terms of a common currency, so that the animal has some means of comparing the 'merits' of sleeping, courting, feeding, etc. In other words, there must be some trade-off process built in to the mechanisms controlling behaviour. McFarland and Sibly (1975) represented the different possible activities as 'candidates' for expression in the behavioural final common path. The strength of candidature, or 'tendency' to perform a particular activity is represented along an axis of a 'candidate space', as illustrated in Figure 1.1. The candidates compete for behavioural expression, and only one candidate can achieve expression at any particular time. Thus the different activities are, by definition, mutually incompatible.

The strength of candidature depends upon the motivational state of the animal, and it is convenient to pinpoint all those states that have the same value in terms of tendency to perform the appropriate

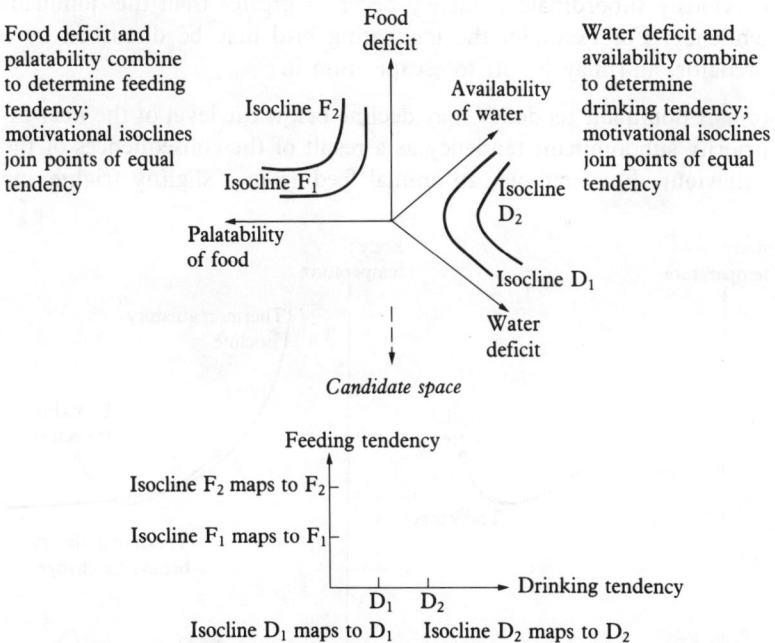

Fig. 1.1 State-space representation of motivation. The motivational space (above) has axes representing the important motivational variables. The motivational state is given by a point in this space, which traces out a trajectory as motivational state changes in time. Lines joining points of equal motivational tendency are called isoclines. These map to points on the axes of the candidate space (below). The position of the point in this space (i.e. the candidate state) determines the animal's behaviour.

3

behaviour, as shown in Figure 1.1. Lines joining these states are called 'motivational isoclines' (McFarland and Sibly, 1975; McFarland and Houston, 1981). Since behaviour itself causes a change in state, it can be represented as a trajectory, which cuts through a number of isoclines, as shown in Figure 1.2. In other words, the predominant tendency changes as a result of performing the behaviour, and its strength may fall below that of a subordinate tendency (i.e. an alternative candidate). At such a point we would expect to see a change in behaviour. According to this simple competition model, changes in behaviour can come about in one of three ways:

(a)   There is a change in internal state such that a previously subordinate tendency becomes greater than the dominant tendency. For example, an incubating bird may become so hungry that it flies off to forage for food.

(b)   There may be a change in the external stimuli such that a previously subordinate tendency becomes greater than the dominant tendency. For example, the incubating bird may be disturbed by a predator, and may fly off to escape from it.

(c)   A dominant tendency may decline below the level of the next-in-priority subdominant tendency as a result of the consequences of the behaviour. For example, an animal feeding in a slightly frightening

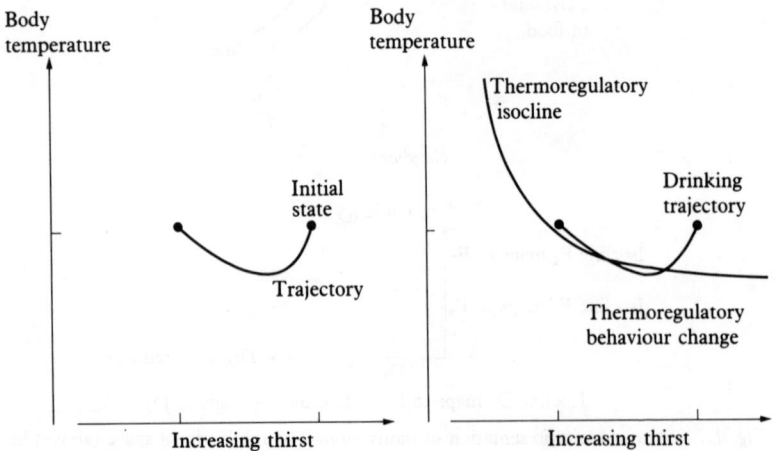

Fig. 1.2 The consequences of behaviour described as a trajectory in motivational space. *Left*: the consequences of drinking cold water (a short-term reduction in body temperature and a long-term reduction in thirst). *Right*: where the trajectory cuts a motivational isocline, changes in behaviour can be expected (from McFarland, 1985).

4

place may become so satiated that its tendency to feed declines below the level of its fear, at which point it moves away from the locality.

The main assumptions behind these conclusions are that the strongest tendency determines the activity that the animal will perform; that all the candidates have equal status in the final common path, and that the choices among candidate tendencies are transitive (see McFarland and Sibly, 1975).

How does this simple decision model shape up to the functional requirements outlined above? The required trade-off among the merits of alternative activities is represented by the motivational isoclines. Differences in the shape and position of isoclines will correspond to differences in the mapping between the motivational state and the corresponding tendency (see Figure 1.1). In general, the cost of being in a particular state will appear as a valuation of internal state variables, while the benefits to be gained by a particular activity will appear as a valuation of the external-stimuli state variables. For example, the tendency to forage for food may be the same in an individual that has a high food deficit (high cost internal state) but low cue strength (low probability of finding food) as in an individual that has a low food deficit but is faced with external stimuli that indicate that there is a good feeding opportunity, as shown in Figure 1.3. This type of trade-off has been widely demonstrated (Sibly and McFarland, 1976; McFarland and Houston, 1981; Stephens and Krebs, 1986).

Having outlined a simple model of motivational competition, we now have to consider whether it will work. There are a number of problems with the model, some of which are discussed in this chapter. The main problem, however is the apparent conflict between the animal's short-term and long-term priorities.

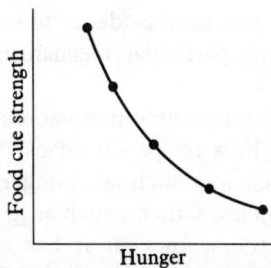

Fig. 1.3 A motivational isocline relating hunger and strength of food cues. The line joins those states (points) that lead to the same strength of feeding tendency (from McFarland, 1985).

Short-term Priorities

An obvious problem with the simple motivational competition model is that the behaviour of the model is prone to 'dithering'. As soon as the model animal changes behaviour, the consequences of the new behaviour will begin to alter the animal's state. This will generally result in a reduction of the dominant tendency. For example, when an animal starts to feed, the first food that it takes reduces its hunger. The reduced tendency is likely to quickly fall below that of the tendency which has just become subordinate, so that the animal reverts to its previous activity. Thus if the animal was nest-building before it started to feed, then it would revert to nest-building soon after commencing feeding.

This problem has long been recognized in motivational modelling, and there have been various suggestions about mechanisms that might prevent dithering (e.g. Ludlow, 1980). The most widely accepted hypothesis is that there is positive feedback from the ongoing activity. Evidence for positive feedback emerged from studies of feeding (Le Magnen, 1985; Wiepkema, 1968) and drinking (McFarland and McFarland, 1968). Food ingested at the beginning of a meal seems to have the effect of strengthening the tendency to feed, and this effect is enhanced by increasing the palatability of the food (McFarland, 1970a; Le Magnen, 1985). Experiments on drinking in doves suggest that oral factors are responsible for a similar type of positive feedback (McFarland, 1965). More recent experiments on incentive point to a similar conclusion (see Toates, 1986, for a review), and it has been argued on theoretical grounds that positive feedback is necessary for behavioural stability (Houston and Sumida, 1985). On the functional side, Larkin and McFarland (1978) pointed out that there is a cost associated with the act of changing from one activity to another, which must be paid as part of the cost of the activity to which the animal is about to change. They produced evidence to support this claim, but they did not postulate any particular mechanism to account for the cost of changing.

Another problem with our simple motivational competition model is that it would seem to be a recipe for chaos. We can imagine that some motivational tendencies, such as hunger, can easily rise to dominate all other tendencies. Others, such as grooming, do not seem to be very good competitors. Indeed, it has often been suggested (Andrew, 1956; Rowell, 1961; McFarland, 1970b) that grooming fits into the gaps between major activities, and is never really a serious competitor. Such an arrangement may be satisfactory when times are easy, but when food is scarce, or when the animal is busy during the reproductive season, candidates which are poor competitors may never achieve dominance in the behavioural final common path. A purely competitive system, with no underlying time allocation principles, would not seem to provide a recipe for an orderly life.

Of course, we may be mistaken in thinking that particular activities are not good competitors. The tendency for grooming, for example, may gradually build up to a point where it can out-compete any other tendency. Some tendencies, such as that of scanning for predators, may rise steeply up to the point where the animal has satisfied itself as to its safety, and then fall quickly. Some of these suggestions may seem rather unlikely, but some such tendencies must occur if competition is to be the only decision-making principle. The alternative is that some other principles are involved, such as preprogrammed scanning for predators, which interrupts whatever the animal is doing. Before addressing this problem we should look more closely at what happens when the animal changes between activities.

Changing from One Activity to Another

When an animal changes from one activity to another it often has to pay for the change, in terms of energy or time. Thus a feeding animal may have to travel to find water to drink, an incubating bird has to travel away from its nest to find food, etc. Moreover, the activity of changing may incur risks from predation. Thus the antelope en route to the water hole is vulnerable to waiting predators, and the bird that quits its hidden nest may thereby reveal the location of the nest. The costs, or decrements in fitness, resulting from time and energy expenditure and risk of predation, are normally offset by the benefits to be gained from the behaviour. Thus the benefits resulting from feeding, drinking and incubation, although offset by the costs, nevertheless make these activities worthwhile. In the case of the activity of changing itself, however, there are no compensating benefits. The result is that when we look at a graph of net costs (Figure 1.4), we find that we cannot follow the normal rule that the animal does that behaviour which brings greatest benefit at each instant of time. If we follow this rule, then the animal will never change to a new activity (where the cost of changing is above zero), because it will always be more beneficial to continue with the current activity rather than take the increase in net cost that changing involves. Similarly, if we assume that the strength of an animal's tendency to do different

7

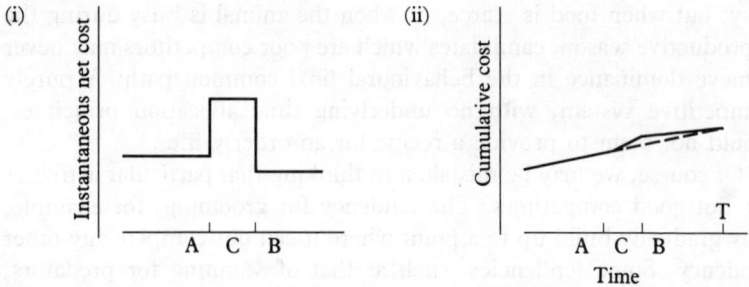

Fig. 1.4 Cost of changing from one activity to another. (i) The instantaneous cost of activity A is higher than that of B, but to change from A to B the animal must do the changing behaviour C, which has an even higher instantaneous cost. (ii) At time T the accumulated cost of the transition A – B – C is the same as it would have been if the animal had remained doing A. At this point it has become worthwhile to have changed from A to B (after Larkin and McFarland, 1978).

activities matches their different net benefits (which is the most simple assumption compatible with motivational competition theory), then we are in a similar conundrum. Thus both functional and mechanistic approaches to the problem of changing from one activity to another involve difficulties. In the functional case, it can be argued that the cost of changing should be added to the cost of the activity that the animal is about to change to (Larkin and McFarland, 1978). This has the effect of postponing the changeover. Empirical studies (McFarland, 1971; Larkin and McFarland, 1978; Larkin, 1981; McFarland, 1983) indicate that changes in behaviour are postponed, or occur less frequently, when the cost of changing is raised.

From the mechanistic point of view, the simple assumption is that the animal changes from one activity to another when the tendency for one activity becomes greater than that for another, for whatever reason. If we compare low and high costs of changing (Figure 1.5), we can see that there are only two possible ways by which postponement could be achieved:

(a) The tendency of the ongoing behaviour could be raised, as in (ii).

(b) The tendency for the subsequent behaviour could be lowered, as in (iii).

The tendency for the ongoing behaviour could be raised as a result of positive feedback (see below), but it is difficult to see how this could be affected by the cost of changing to some future activity. The

8

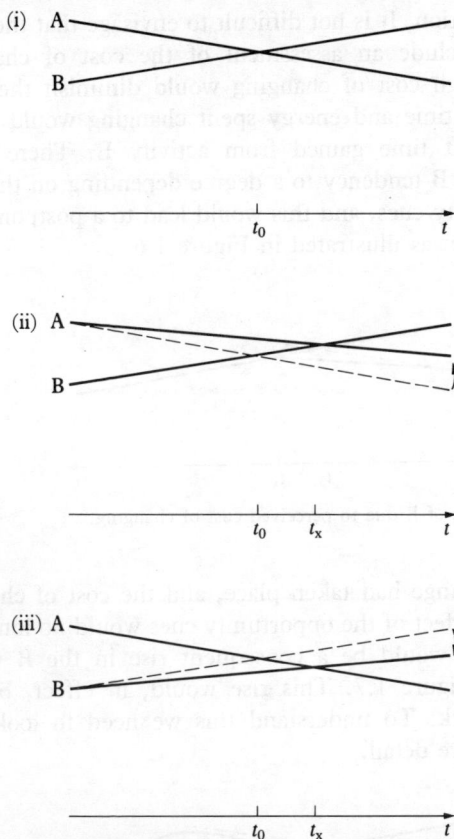

Fig. 1.5 Changes from one activity to another occur when the benefits of the ongoing activity (A) become less than those of an alternative activity (B). (i) Zero cost of changing (observe change from activity A to activity B at t_o). (ii) Cost of changing = x results in raising A (observe change in activity at t_∞). (iii) Cost of changing = x results in lowering B (observe change in activity at t_x).

tendency for the subsequent behaviour could be lowered as a result of the animal's perception of the situation, and this could be affected by the cost of changing.

Suppose that an animal engaged in activity A, is likely to change to activity B. The strength of the tendency for B will depend upon the internal causal factors influencing B in (some) combination with the cue strength (the animal's assessment of the external factors) for B. The cue strength will be (designed to be) influenced by cues indicating the opportunity value (e.g. the likelihood of obtaining food/unit time, if the animal were to spend time foraging) of the activity in question. It is not difficult to envisage that the opportunity value could include an assessment of the cost of changing. Cues indicating a high cost of changing would diminish the opportunity value, because time and energy spent changing would diminish the net returns/unit time gained from activity B. There would be a lowering of the B tendency to a degree depending on the strength of these opportunity cues, and this would lead to a postponement of the A-B changeover, as illustrated in Figure 1.6.

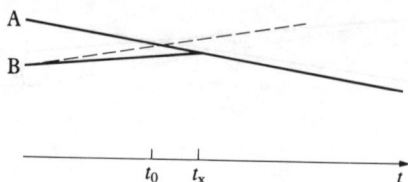

Fig. 1.6 Lowering of B due to perceived cost of changing.

Once the change had taken place, and the cost of changing paid, the inhibiting effect of the opportunity cues would no longer be effective, and there would be a consequent rise in the B tendency, as illustrated in Figure 1.7. This rise would, in effect, be a form of positive feedback. To understand this we need to look at positive feedback in more detail.

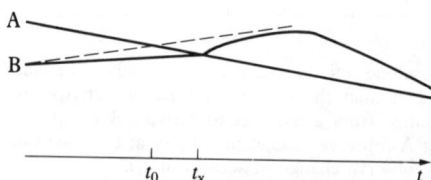

Fig. 1.7 Rebound of B due to removal of perceived cost of changing.

The first definite proposal was by McFarland and McFarland (1968) who found it necessary to introduce positive feedback into their control model of drinking in doves (Figure 1.8), not to prevent dithering between activities (they were considering only one activity), but to account for the fact that drinking after interruption is resumed at a lower rate than that at which it was terminated (Figure 1.9). As we have seen (above), various authors have suggested that such positive feedback would prevent dithering, but this was first explicitly shown by Houston and Sumida (1985).

10

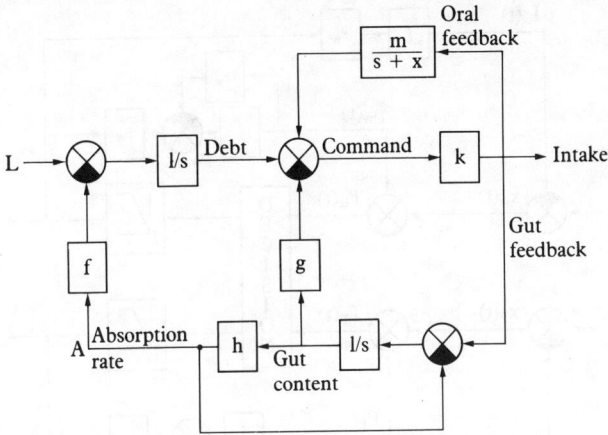

Fig. 1.8 Control model of drinking in doves, in which the oral factors provide positive feedback and the gut factors provide negative feedback (the circles containing X are summing points, in which a black quadrant changes the sign). L = rate of water loss, A = rate of absorption into the blood. The other symbols are the parameters of the component mechanisms (after McFarland and McFarland, 1968).

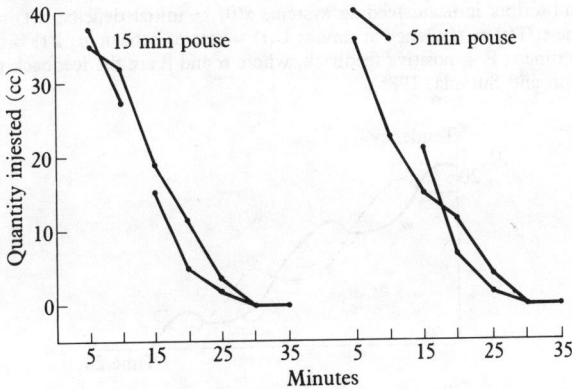

Fig. 1.9 Rate of drinking under normal conditions (continuous curve) and when interrupted for 5 or 15 min (broken curve). The break in the curve occurs at the point at which drinking was interrupted, but the size of the gap does not indicate the period of the pause. Results averaged over five subjects (after McFarland and McFarland, 1968).

Houston and Sumida (1985) investigated the behaviour of two coupled systems, which they called the feeding system and the drinking system, each similar to the model proposed by McFarland and McFarland (1968). Their model is illustrated in Figure 1.10. If this model decided between feeding and drinking on the basis of the

11

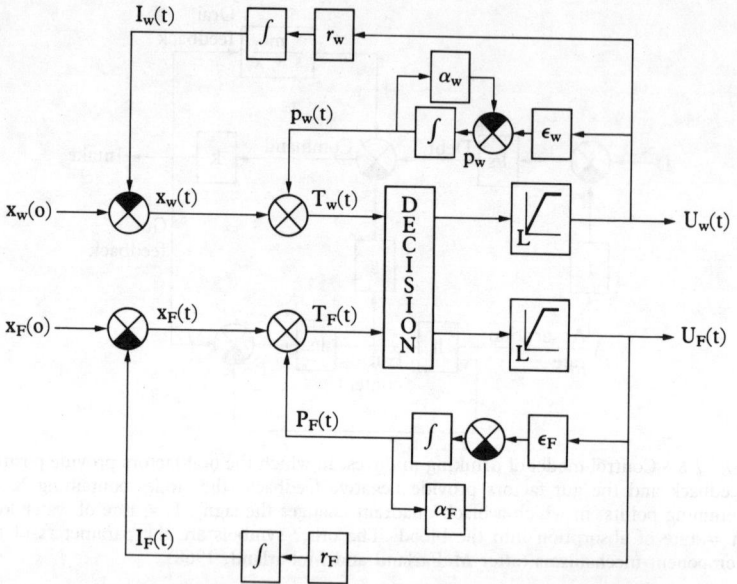

Fig. 1.10 Block diagram of positive feedback model. W subscripts indicate drinking system. F subscripts indicate feeding system. $x(0)$ = initial deficits; $x(t)$ = value of deficit at time t; $T(t)$ = tendency at time t; $U(t)$ = rate of behaviour; $I(t)$ = cumulative intake up to time t; P = positive feedback, where α and β are the feedback parameters (after Houston and Sumida, 1985).

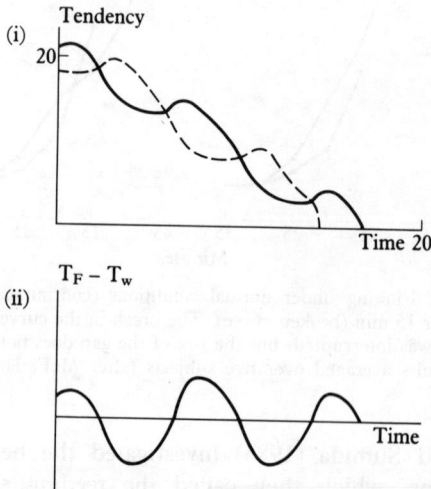

Fig. 1.11 (i) Simulated dynamics of two tendencies versus time. Following a switch there is an initial rise in the tendency of the new activity because of positive feedback. Solid line = feeding tendency. Broken line = drinking tendency. (ii) Difference of two tendencies versus time (after Houston and Sumida, 1985).

food or water deficits (x_F and x_W), it would either jam or dither. To produce bouts of feeding and drinking each system has a positive feedback loop, which boosted the tendency for the activity that was being performed. The tendency for each activity is simply its current deficit plus the current value of its positive feedback.

The boost in tendency provided by the positive feedback occurs at the beginning of a feeding or drinking bout, as shown in Figure 1.11. The effect of this is to induce an initial separation between the two tendencies, which then decays under the influence of negative feedback from the consequences of the behaviour. The result is that the behaviour sequence is broken up into bouts, the size of which is determined by various parameters of the model (Houston and Sumida, 1985).

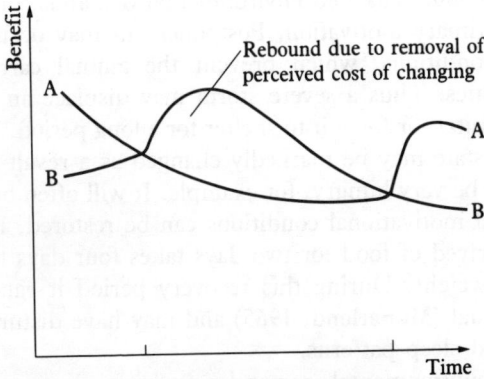

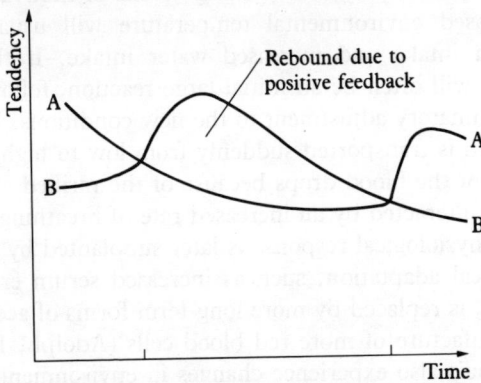

(i) Functional interpretation (from Fig. 1.7)

(ii) Mechanistic interpretation (from Fig. 1.11)

Fig. 1.12 Comparison of functional (i) and mechanistic (ii) models of changing from one activity to another.

The similarity between the cost of the changing model (Figure 1.7) and the Houston and Sumida model (Figure 1.11) can be seen in Figure 1.12. These two models are complementary, since the former specifies the mechanism (in terms of opportunity cues, etc.) while the latter specifies the dynamics (in terms of control formulae). It is quite possible that the formulae describe the mechanism. What is needed is a test of the implication that the magnitude of the positive feedback is proportional to the cost of changing.

Long-term Priorities

Long-term motivational priorities may arise as a result of postponement of behaviour, changed environmental demands, or endogenous changes in primary motivation. Postponement may occur as a result of adverse conditions, which prevent the animal carrying out its normal activities. Thus a severe storm may displace an animal from its normal habitat, or force it to shelter for a long period. The animal's motivational state may be markedly changed as a result of postponement. It may be very hungry, for example. It will often be a long time before normal motivational conditions can be restored. For instance, a pigeon deprived of food for two days takes four days to recover its normal bodyweight. During this recovery period it eats and drinks more than usual (McFarland, 1965) and may have disturbed thermoregulatory and sleep patterns.

Changed environmental demands, such as changes in environmental temperature, may induce many weeks of motivational adjustment. Increased environmental temperature will usually result in reduced food intake and increased water intake. If the change is sudden there will often be an initial large reaction, followed by three weeks of acclimatory adjustment to the new conditions. For example, when a person is transported suddenly from low to high altitude, the oxygen level of the blood drops because of the rarified air. This drop is initially counteracted by an increased rate of breathing. This rapid, and costly, physiological response is later supplanted by slower forms of physiological adaptation, such as increased serum erythropoietin. This, in turn, is replaced by more long-term forms of acclimatization, like the manufacture of more red blood cells (Adolph, 1972).

Most animals also experience changes in environmental conditions between night and day. In addition to direct effects of changes in the weather, such as a drop in temperature, there are indirect effects, such as alterations in food availability, and in numbers of predators, brought about by changes in temperature, light intensity, etc. In adjusting to the differences between night and day, the animal adopts a daily routine which is made up of many different aspects of behaviour fitted together to form a pattern that tends to be repeated day after day, provided the prevailing conditions do not change much from one day to the next. When a single activity is studied it tends to follow a typical daily rhythm, usually called a circadian rhythm. Such rhythms have received considerable attention from those primarily interested in circadian clocks (Rusak, 1981), but the daily routines of animals have been subjected to relatively little study (Daan, 1981).

From the functional point of view, it is not surprising that rhythms of rest and activity are widespread in the animal kingdom. When it is disadvantageous to be active at night, the best policy is to sit tight in a safe place and save as much energy as possible. This has been suggested as one of the prime functions of sleep (Meddis, 1975). Nocturnal species may hide during the day if they are likely to be preyed upon. If they are themselves nocturnal predators, they may remain hidden and inactive during the day to avoid scaring their prey. Thus the daily rhythms of the physical environment make some activities advantageous at one time and disadvantageous at another. Much depends upon the ecological niche of the species concerned.

From the functional viewpoint, we would expect that an animal which is adapted to its environment will settle into a daily routine designed by natural selection to maximize the survival value of its various activities. This is partly a matter of making the most of opportunities. For example, the European kestrel specializes in preying upon small mammals. The bird relies on vision to detect and catch its prey and hunts most successfully when the light is good. Field studies (Daan, 1981) reveal that kestrels catch prey throughout the day but do not always immediately eat what they catch. They tend to catch surplus prey in randomly chosen locations in their hunting area. The catching occurs throughout the day and the cached items are typically retrieved at dusk. This routine enables the kestrel to make the most of the available prey during the day, without spending too much time eating the prey. If the bird ate all its surplus prey immediately it would become unnecessarily heavy and its hunting efficiency would probably decline.

In addition to their species' typical daily routines, individual animals may develop their own daily habits. Thus kestrels that have found food at a particular time and place tend to repeat the same search pattern the next day. Such a strategy is appropriate in circumstances where the prey also has its typical daily routine. Field observations such as these can provide useful circumstantial evidence about the daily routines of animals, and can suggest certain types of functional interpretation. To carry the investigation further, however, we need to relate these observations to a consistent body of theory. In recent years attempts have been made to relate the temporal organization of behaviour to the theory of natural selection.

In considering the medium-term temporal organization of behaviour the most obvious design criteria are: (i) minimizing risk and (ii) maximizing opportunities, and neglecting no important behaviour. These can be boiled down to the more fundamental criteria of maintaining physiological stability, avoiding predators and maximizing reproductive success (McFarland and Houston, 1981). Some of these points can be illustrated by a recent study of herring gulls.

Consider the situation of a herring gull incubating its eggs. Normally, both parents incubate in turn and a nest left unattended is soon subject to predation by neighbours. A sitting bird will not quit the nest until relieved by its partner, unless it is flushed from the nest by a predator, such as a fox or human. Even then it remains in the vicinity, and often attacks the intruder. Usually the partner leaves the nest to forage for food and returns to the territory within a few hours. Sometimes, however, the return is delayed as a result of some mishap, injury, or capture by a scientist. The sitting bird cannot feed, and its increasing hunger should eventually cause it to desert the nest. Herring gulls breed in many successive seasons and it is not in the genetic interests of an individual to endanger its life for a single clutch of three eggs, which are likely to give rise to only 0.8 fledglings.

The basic dilemma for the incubating bird lies between the possibility that its partner may return at any time, and the increasing necessity to obtain food. The situation is complicated by the fact that the gull must assess the return on time invested in foraging, which may vary considerably with the weather, the state of the tide, and the time of day. Moreover, individual gulls often specialize in particular types of foraging, so that the sitting bird should take account of its own foraging skills in assessing the likely returns from the time spent foraging in different places. It may seem that the best course of action requires quite a bit of computation on the part of our herring gull. I am not suggesting, however, that this is done cognitively. Rather, it is more likely that the bird relies on rules of thumb, based upon its normal daily routine and habitual behaviour.

Sibly and McCleery (1985) have conducted a sophisticated functional analysis of the herring gulls at Walney during the incubation period. They describe a procedure for testing whether a particular set of decision rules maximizes fitness and they use it to identify optimal decision rules governing feeding, incubation, and the presence on territory of herring gulls during incubation. The characteristics of the optimal decision rules depend upon the risks attendant upon each aspect of behaviour. Therefore in calculating the rules it is necessary to measure the risks an animal takes in carrying out its various activities. Using such estimates they identify the characteristics of the optimal decision rules and use them as a basis for evaluating the

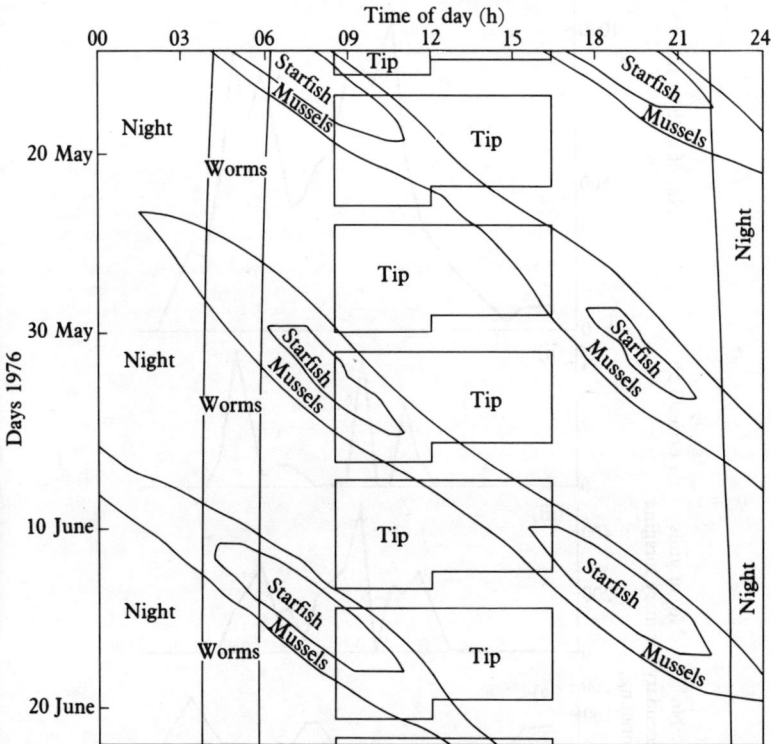

Fig. 1.13 The availability of the major foods of herring gulls at Walney in 1977. 'Worms' indicates the availability of invertebrates (mainly earthworms) on pasture fields on the mainland within an hour of sunrise; 'Tip' indicates the availability of domestic waste at refuse tips 08.30–16.30 Monday to Friday, 08.30–12.00 on Saturdays; 'Mussels' indicates the availability of first year mussels and other invertebrates on mussel skears when the tide was below 1.9 m below O.D.; 'Starfish' indicates the availability of starfish and other invertebrates on mussel skears when the tide was below 3.1 m below O.D.; 'Night' was from one hour after sunset to one hour before sunrise, when food was assumed to be unavailable (from Sibly and McCleery, 1985).

17

fitness of the observed behaviour. The method is a development of that first proposed by Sibly and McFarland (1976).

The main risks of reproductive failure during the incubation period are of egg death due to exposure or predation, and of parent death as a result of accident, disease, predation or starvation. These risks can be quantitatively evaluated by experiment (Sibly and McCleery, 1985). The opportunities relate mainly to feeding. Figure 1.13 shows the availability of the major foods taken by herring gulls at Walney.

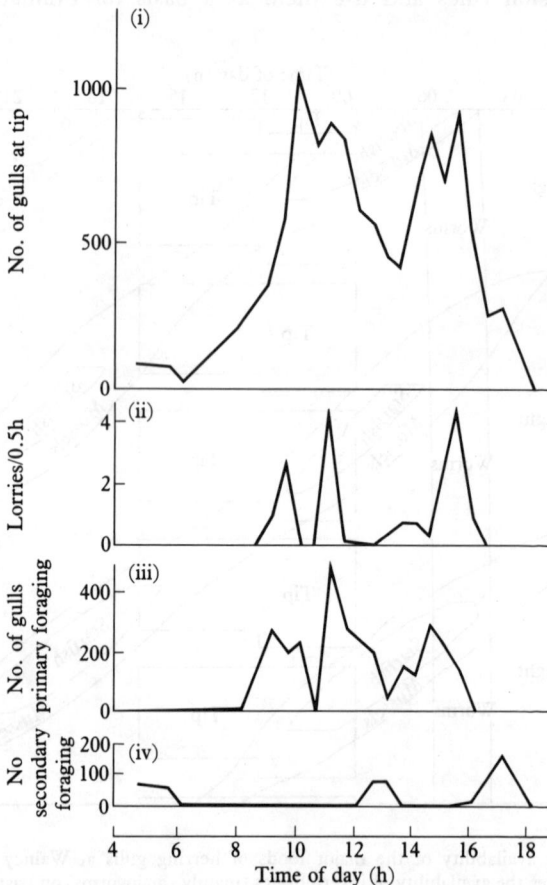

Fig.1.14 Gull activity at Walney refuse tip on weekdays, April–June 1976. About 90 per cent were adult herring gulls, 10 per cent adult lesser black-backs. (i) Average number of gulls at the tip, in adjoining fields, or on the adjoining beach. Each data point is based on the average of four observations. Band of grey indicates ± 1 SD. (ii) Number of lorries carrying domestic waste that unloaded in half an hour. (iii) Average number of gulls engaged in primary foraging. (iv) Average number of gulls engaged in secondary foraging (from Sibly and McCleery, 1983).

The various colony members specialize (to the extent of having preferred sources of food) in feeding on different rhythmically available resources. These include refuse and earthworms, available on a circadian basis, and mussels and starfish available on a tidal cycle. Each gull has to adjust its temporal pattern of incubation to permit both itself and its partner to take advantage of the feeding opportunity profiles that each is specialized in exploiting, without leaving the nest unattended. An opportunity profile is a function of time that represents the returns that the animal would obtain if it foraged at that time. An example is illustrated in Figure 1.14. Successful incubation by a breeding pair requires complex co-operation and behavioural synchronization with each other and prevailing circadian, tidal, and social rhythms.

Sibly and McCleery (1985) found that the feeding preferences of individual gulls were very important determinants of fitness. They constructed a fitness landscape for each breeding pair (an example is shown in Figure 1.15), based on two important hunger variables, which they called normal feeding threshold and desperation feeding threshold. The former is the degree of hunger at which a bird would

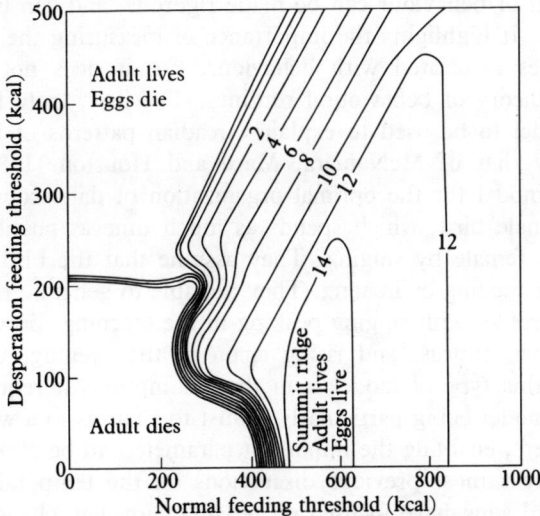

Fig. 1.15 A selective landscape with desperation feeding threshold and normal feeding threshold as axes. The lines represent contours of equal fitness. There are three main features of this landscape. Maximum fitness is attained along the 'summit ridge', where the adult lives and the eggs hatch. To the right of the ridge, fitness declines because of the cost of carrying extra weight. Minimum fitness occurs when the adult dies. On the plateau at top left the adult lives but the eggs die (from Sibly and McCleery, 1983).

leave the territory to forage if one of its preferred foods became available, provided that its partner was on territory to take over incubation. The latter is the degree of hunger at which the bird would leave its nest to feed when a suitable opportunity arose, even if its mate was not available for incubation duty. On the basis of the fitness calculations Sibly and McCleery (1985) identified various characteristics of the optimal behaviour strategies. In particular, they predicted:

(a)   Energy reserves should be maintained between 500 and 1200 kcal.

(b)   The members of a mated pair should have complementary food preferences, with at least one being a preference for feeding at the local refuse tip.

(c)   The parent should desert the offspring if the energy reserves fall below 200 kcal.

They were able to test these predictions by observation and experiment.

This example show how functional theories about the temporal organization of behaviour can be made rigorous, and can be empirically tested. It highlights the importance of measuring the risks and opportunities associated with behaviour, but it does not, strictly, provide a theory of behavioural routines. The first, truly functional formal model to be used to explain circadian patterns of behaviour is probably that of McNamara, Mace and Houston (1987). They develop a model for the optimal organization of daily routines in a territorial male bird, which spends as much time as possible trying to attract a female by singing. They assume that the bird's time is spent either feeding or in song. They are able to generate the typical passerine profile, with singing peaking in the morning, providing the familiar dawn chorus, and rising again in the evening. Our main interest is this type of model is in the assumptions it requires, this particular model being particularly robust to changes in a wide range of parameters, enabling the important parameters to be clearly identified. Whereas most previous discussions on the temporal organization of bird song have focused on possible circadian changes in the profitability of singing or foraging, this model suggests that such considerations are not so important as was thought. This type of functional analysis illustrates how the factors involved in the organization of daily routines are not always intuitively obvious, but may be brought to light by a formal model. It is this aspect of functional analysis which is especially important for our understanding of the mechanisms underlying daily routines.

We now turn to the third of the design criteria mentioned earlier, that of neglecting no important behaviour. The question is whether the animal has sufficient time do all those activities that are important for survival and reproduction. In herring gulls, we can expect that behaviour that has the prime function of promoting the survival of the individual will take precedence over behaviour that promotes other aspects of fitness, such as territorial mating and parental behaviour. However, species vary in this respect. Some aspects of behaviour are essential, but others like thermoregulatory behaviour may be important only when physiological mechanisms cannot cope. Thus, drinking is a daily necessity for some species, but other species can manage without drinking at all. Each activity has value in terms of fitness, and animals must be designed to allocate priorities to activities in a general way, as well as from minute to minute.

We can approach this problem by considering a measure of the cost to an animal of abstaining from each activity in its natural repertoire (Houston and McFarland, 1980). If an animal did no feeding, for instance, the cost would be high, but if it abstained from grooming the cost might be relatively low. An animal that had a high tendency to both feed and groom but that did not have time to do both would sacrifice less, in terms of fitness, if it devoted its available time to feeding. Suppose an animal fills its typical day with useful activities, as illustrated in Figure 1.16. In an environment that was much the same from day to day, the animal would adjust its activities to the time available. Suppose, however, there is a change in the environment such that it now takes very much longer to obtain the normal amount of food (see Figure 1.17). The animal can respond to the changed circumstances by spending the same amount of time feeding

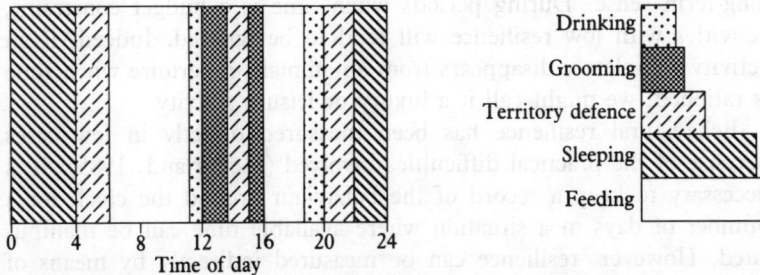

Fig. 1.16 Diagram showing how an animal might spend its time throughout the day (from McFarland, 1985).

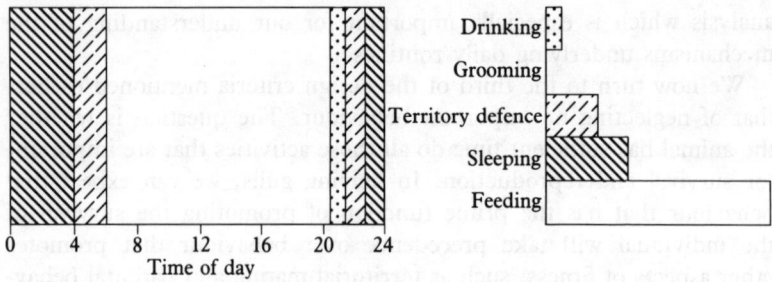

Fig. 1.17 Diagram showing how the animal in Figure 1.16 might adjust its daily routine when required to spend much more time to obtain the same amount of food (from McFarland, 1985).

as before and settling for less food. Alternatively, it could insist on the same amount of food as usual, or it could compromise between the two extremes. If the animal spent a long time obtaining the usual amount of food, then there would be less time available for all the other activities in its repertoire. These would have to be squashed into the remaining time. The extent to which an activity resists squashing can be represented by a single parameter, which we called resilience. In the case where the animal feeds for the normal amount of time and ends up with a reduced food intake (Figure 1.16), the resilience of feeding is relatively low because feeding has not compressed the other activities even though the animal's hunger is increased. In the case where the animal insists on the normal amount of food (Figure 1.17), the resilience of feeding is relatively high because feeding ousts the other activities from the time available without itself being curtailed in any way.

Behavioural resilience is a measure of the extent to which each activity can be squashed in terms of time by other activities in the animal's repertoire. It also reflects the importance of an activity in a long-term sense. During periods when time is a budget constraint, activities with low resilience will tend to be ignored. Indeed, if an activity completely disappears from an animal's repertoire when time is rationed, we might call it a luxury or leisure activity.

Behavioural resilience has been measured directly in few cases because of the practical difficulties involved (McFarland, 1985). It is necessary to keep a record of the behaviour around the clock for a number of days in a situation where available time can be manipulated. However, resilience can be measured indirectly by means of demand functions. These are used by economists to express the relationship between the price and the consumption of a commodity.

For example, when the price of coffee is increased in the supermarket, people continue to buy about the same amount as before, perhaps a little less. As the price of fruit is increased, however, the demand for fruit falls off. When the price of fresh fish is increased, demand declines markedly. Presumably, people are willing to pay more to maintain their normal coffee-drinking habits. Demand for coffee is said to be inelastic. If the price of fresh fish increases, however, people tend to buy less and to switch to substitute foods such as meat or canned fish. Demand for fish is said to be elastic.

Exactly analogous phenomena occur in animal behaviour. If an animal expends a certain amount of energy in a particular activity, then it usually does less of that activity if the energy requirement is increased. Numerous studies have shown that demand functions in animals follow the same general pattern as those of humans (Lea, 1978). For example, S Lea and T Roper (1977) found that demand for food was inelastic in rats required to work (press a lever) to obtain food rewards. As the work required (lever presses per reward) increased, the rats continued to work for about the same amount of food. However, these investigators also found that elasticity of demand for food pellets increased when sucrose was available as a substitute.

The parallel between the demand phenomena of animals and people is a topic of considerable current interest (e.g. Allison, 1979, 1983; Lea, 1978; Rachlin, 1980) (see also Chapters 2 and 3 of this volume). The elasticity of demand functions gives an indication of the relative importance of the commodities (or activities) on which the person (or animal) spend his or her money (or energy). There is a close relationship between elasticity of demand and resilience (Houston and McFarland, 1980). Thus, demand functions can be used as indirect measure of resilience. If activity A has higher resilience than activity B, then A will tend to show a relatively inelastic demand function and B will show an elastic one (McFarland and Houston, 1981).

There is little theoretical understanding of what happens when an important part of an animal's normal daily activity takes a much shorter time that usual. This happens to animals in captivity that cannot indulge in their normal range of activities and are handed their food on a plate (see Chapter 2). There have been some experimental attempts to manipulate the daily routines of herring gulls in the wild. Niebuhr (1981) placed feeding boxes on the territories of incubating gulls at Walney. Each feeding box contained a circular tray holding six cups which were filled with a total of 200 g of cat food twice daily, at 10.00 hours and 16.00 hours. Each box had a gold-coloured key which, when depressed, caused the circular tray within the box to move around one position so that the next cup of food became available through a hole in the lid of the box. The gulls soon learned to operate the mechanism to obtain food. Control pairs of gulls were provided with replica boxes containing no food. Compared with the controls, the experimental birds spent much more time on the territory and their incubation schedules became desynchronized from the foraging opportunity profiles. The experimental birds were able to obtain their normal daily food requirement from the boxes, and they seldom foraged. In a similar study, Shaffery, Ball and Amlaner (1985) found that the extra time spent on the territory did not result in an increase in total time spent sleeping, but did lead to a dissociation between two types of sleep. They suggested that one of these (front-sleep) was normally a consequence of foraging, and was infrequent in the experimental birds because they did not have to forage. These studies throw some light on the problem of leisure (see Chapter 2), but they do not greatly add to our theoretical understanding of resilience.

The main contribution of functional analysis to our understanding of the temporal organization of behaviour can be summarized as follows. We can calculate the optimal temporal strategy in particular situations, and compare this with the observed behaviour of the animals concerned (Sibly and McFarland, 1976; McCleery, 1978; McFarland and Houston, 1981; Krebs and McCleery, 1984). We can see that the cyclic nature of environmental changes will mean that the optimal strategy on successive days is similar. In particular, circadian fluctuations in risks and opportunities will result in optimal temporal strategies that are similar on successive days. In an environment that changes little from day to day, there will be an optimal daily routine which will reflect the circadian risks and opportunities to some extent, but will also be affected by the resilience of the prevalent activities.

Underlying Mechanisms

The daily routines of animals result from the combination of exogenous and endogenous influences. The exogenous influences are the external stimuli that influence behaviour, especially ambient temperature, light intensity etc., which change on a circadian basis. We should include among these the influence of the animal's endogenous circadian clock, which provides time cues that play a role similar to that of external cues (see McFarland and Houston, 1981 (p. 44), for a discussion of this point). The endogenous influences result largely from the mechanisms governing the interactions between motivational systems, and from the endogenous aspects of the internal clock, such as circadian hormone release, etc.

Three main levels of interaction between motivational systems are distinguished in Figure 1.18. These have been called the primary level, the secondary level, and the level of the behavioural final common path (McFarland, 1971, 1978). At the primary level, influences may come directly from the environment, as instanced by the effect of ambient temperature upon the thermoregulatory system, but more commonly they come from other motivational systems. Thus the effects of ambient temperature upon feeding and drinking operate via the thermoregulatory system. Figure 1.18 shows the four main ways in which such interactions occur:

(a) Physiological changes in one system may result from behavioural changes in another system. Thus increased water loss can be a direct consequence of food intake.

(b) Physiological changes in one system may result from changes of motivation in another system. Thus a particular motivational state can alter the hormonal balance. At the secondary level there are also two main types of interaction.

(c) There may be changes in motivational state in one system consequent upon behaviour in another system. Thus the condition of

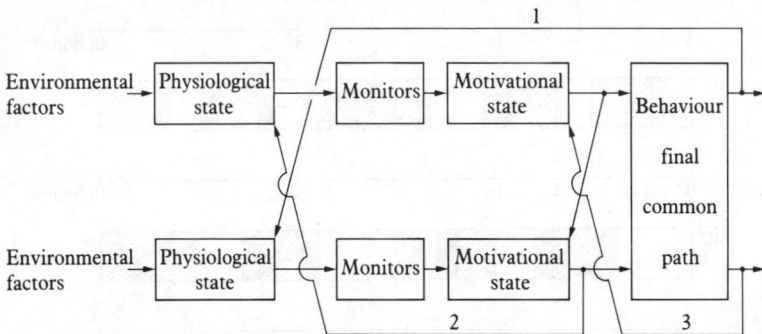

Fig. 1.18 Generalized representation of the possible types of interaction between motivational systems. Feedback loops within systems have been omitted for clarity. The numbers 1–4 refer to the types of interaction mentioned in the text (from McFarland, 1978).

25

a bird's nest, and the presence of eggs in the nest, may cause shifts in motivational state which make the animal more likely to incubate.

(d)   There may be direct interaction between motivational states in different systems. Thus thirst has an inhibitory effect on eating.

Examples of various types of motivational interaction are discussed by McFarland (1971, 1978) and Toates (1980, 1986).

The third level of interaction, the behavioural final common path (McFarland and Sibly, 1975) is essentially the mechanism by which the animal decides among incompatible behavioural tendencies. They assumed that the strongest tendency wins, and is expressed as overt behaviour, while the behavioural expression of all the other activities is inhibited. In other words, the animal can engage in only one activity at a time.

The effects of motivational interactions upon the temporal organization of behaviour can be seen by considering what would happen in a completely unchanging environment. Assuming that the system had reached steady-state, then a pattern of behaviour would emerge that would be repeated indefinitely, as illustrated in Figure 1.19(i). The period of repetition would depend upon the natural frequency of the system, taken as a whole (McFarland, 1971). This would depend upon the characteristics of the motivational interactions, and of the feedback systems within each motivational system. If the

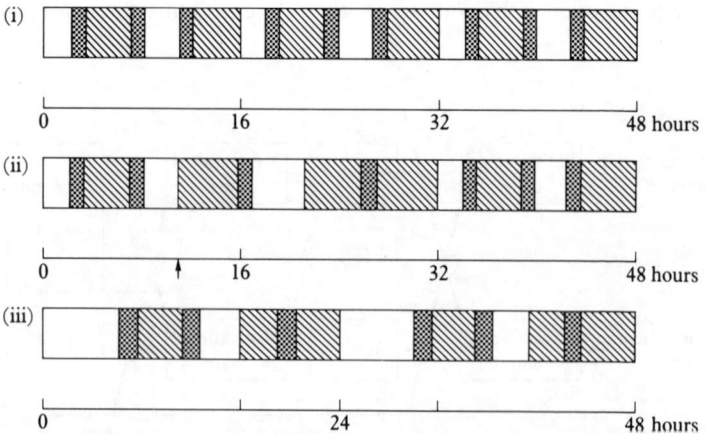

Fig. 1.19   Rhythm in behaviour sequences. (i) Inherent rhythm produces a cycle of period 16 hours in an unchanging environment. (ii) Transient changes in (i) due to a disturbance (arrow). (iii) Circadian rhythm due to cyclic changes in the environment, or to an endogenous circadian clock.

system were subject to a step-change in environmental conditions, then there would be a transient period of adjustment, before the behaviour settled down to a new steady-state pattern, as illustrated in Figure 1.19(ii).

Suppose we now introduce a circadian element: either some influence of the endogenous circadian clock, or an exogenous circadian influence. This will be equivalent to driving the system at a given frequency (McFarland, 1971) and the behaviour will inevitably take up a circadian pattern. Even if the circadian influence consisted in a single event every 24 hours, a daily routine would result. We can see that this is so if we consider an animal that is immobilized during darkness, but wakes each day into an unchanging environment, as shown in Figure 1.19(iii). The awakened animal will have simultaneous tendencies for a number of different activities, but only one will initially be expressed as overt behaviour. Let this first activity be feeding. This will have two types of consequence:

(a) There will be consequences resulting from the postponement of the behavioural expression of other systems.

(b) There will be consequences that affect both the feeding system, and other systems, resulting from the behavioural, psychological and physiological effects of feeding. These consequences will cause some physiological changes and some consequent realignment of the motivational priorities.

Sooner or later, however, another activity will take over from feeding, as a result of a decline in the feeding tendency due to satiation, or as a result of an increase in some other tendency, or a combination of both. The main characteristic of an animal that has

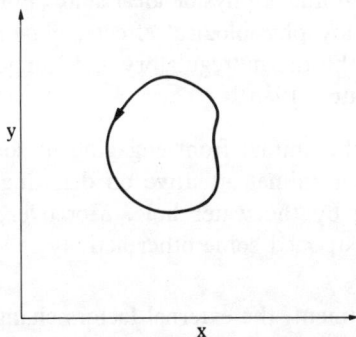

Fig. 1.20 Closed trajectory in state space due to repeated daily routines.

a settled daily routine is that it will have a closed trajectory in the state space as shown in Figure 1.20. In other words, the state of the animal will tend to be the same at each particular time of day. Our first task is to identify the mechanisms which contribute to this trajectory, so that we can gain some idea of what will happen when the daily routine is disturbed.

The motivational state of an animal at any particular time depends upon: (i) its physiological state, (ii) the cue state arising from its perception of the external world and (iii) the consequences of its current behaviour. The relationships among these factors have been discussed at some length elsewhere (McFarland and Sibly, 1972; Sibly and McFarland, 1974; McFarland and Houston, 1981). (For non-technical accounts see Huntingford, 1984; Halliday and Slater, 1983; McFarland, 1985.) Basically the state at any particular time determines which behaviour occurs. The state changes continually as a result of the animal's own behaviour, and this can be described as a trajectory in a state-space, as illustrated in Figure 1.2.

The behaviour of the animal has consequences which depend upon the situation in which the behaviour occurs. These consequences influence the animal's motivational state in three main ways:

(a) By altering the external stimuli perceived by the animal, and hence altering the cue state. For example, in drinking at a water hole an animal may (typically) encounter other members of its species, the perception of which alters its state.

(b) By direct feedback onto the motivational state. For example, the act of drinking may have short-term satiation effects that are independent of water injestion (Rolls and Rolls, 1982).

(c) By altering the animal's physiological state. For example, ingestion of water has many physiological effects, not only on the thirst system, but also on the thermoregulatory and hunger systems (Rolls and Rolls, 1982; Toates, 1986).

(d) By preventing the animal from engaging in some other behaviour. For example, an animal attentive on drinking may not notice the predator lurking by the water hole. Moreover, by visiting the water hole, it has postponed some other activity.

In a stable environment, the external factors change on a circadian basis in a manner which is similar from one day to the next. There will, of course, be random fluctuations in local conditions, but these will tend to be ironed out by the animal's internal compensating mechanisms (see below). Thus, once an animal has settled down in a stable environment, the temporal pattern of its behaviour will be much the same from day to day. Because the situation in which the behaviour occurs will be much the same from day to day, though it will change on a circadian basis, it follows that the consequences of the behaviour will be much the same from day to day, though changing throughout the day. If the external stimuli and the consequences of behaviour follow the same pattern on successive days, then the state trajectory (the change in state with time) will be similar on successive days. In other words, the trajectory will be a closed one, as illustrated in Figure 1.20.

It is worth noting, however, that we would expect to see some kind of daily routine even under constant non-cyclic external conditions. Suppose our animal is predisposed to sleep at night, and we allow some slight *zeitgeber* (Aschoff, 1960) to entrain the circadian clock. When the animal awakes it will not have eaten for a certain amount of time and will be somewhat hungry. If it spends the next period of time foraging, it will postpone other activities even more, because it cannot simultaneously forage and engage in other activities. We have, in effect, introduced a step input into a multi-output system that has an incompatibility constraint (the behavioural final common path). In the absence of any further input, there will be a succession of activities during a transient period, followed by a steady-state routine, as illustrated in Figure 1.19. If the step is repeated every 24 hours (due to the *zeitgeber*), the transient will be repeated and a daily routine will be the result.

Small disturbances in the daily routine will tend to be self-correcting for two main reasons:

(a) The trajectory is locally stable due to the quick-acting regulatory mechanisms that are able to compensate for small deviations in the state (Sibly and McFarland, 1974).

(b) Small deviations in the trajectory will not generally induce major deviations in behaviour, because there is a low probability of the state moving into a region of the space responsible for a different activity. A large disturbance in the daily routine can have a destabilizing effect if it causes major changes in the consequences of behaviour, if it induces physiological acclimatization, or if it causes the animal to learn a new pattern of behaviour.

Disturbance of the daily routine can come about as a result of:

(a) Changes in the external stimuli which alter the animal's cue state.

(b) Changes in the environmental situation which alter the consequences of behaviour.

(c) Changes in the animal's internal state which result from endogenous factors such as seasonal hormonal changes.

I am primarily concerned here with the first two of these possibilities, since the third is basically a pre-programmed change in the overall situation.

Changes in external stimuli that are likely to induce changes in daily routines would include a new source of food, a possible new indication of danger, etc. The extent to which the animal responds to such stimuli will depend upon the exact situation. There are certain processes which will tend to diminish the animal's response, such as neophobia and habituation. Many animals avoid novel situations, or treat them cautiously. This phenomenon has been well documented in the investigations of bait shyness in rats (Rozin, 1976b). If potential signs of danger are repeated without apparent deleterious consequences, then the animal habituates, and its response to the disturbance progressively diminishes through habituation. Often the effects of the disturbance will be profound, and an animal may radically alter its feeding routine to exploit a new resource. In response to danger the animal may alter its feeding habits, or even move to a new habitat.

Similar considerations apply to changes in the consequences of behaviour that result from alterations in the environmental situation. For example, there may be a change in the availability of food that is not immediately apparent to the animal, but which results in an increase in the time spent foraging. This will mean that less time is spent on other activities. Of course, the animal will soon discover that the consequences in its behaviour have altered, and will make a number of adjustments.

In considering the adaptation of individual animals to changes in the environment we have to take account of: (i) the extent to which the animal is designed to cope with change, and (ii) the ways in which the animal detects the environmental change. Animals that have evolved in stable, unchanging environments are likely to make decisions on the basis of relatively stereotyped sets of rules tailored to suit the ecological niche. Many species show genetically preprogrammed forms of adaptation, both acclimatization and learning. Some show acclimatization changes on a seasonal basis, often governed by an endogenous circannual clock. The physiological and behavioural changes that anticipate hibernation and migration come into this category (see McFarland, 1985 for a review). Preprogrammed forms of learning are particularly important in the development of the individual, but are not particularly relevant to daily routines. There are three main mechanisms by which animals may adapt to environmental change in a preprogrammed manner (McFarland and Houston, 1981):

(a) Changes in the animal's decision rules may be cued by habitat factors. In other words, when in one habitat the animal uses one set of rules, when in another it uses another set. Such contingency planning is obviously important in migratory animals, and may also occur in domestic animals which move from one habitat to another (e.g. indoor-outdoors) every day.

(b) There may be motivational changes that are directly cued by external stimuli. Thus an animal may become more wary in a place associated with danger, or by signs of danger, such as the sight of a particular person.

(c) There may be basic physiological changes. For example, the critical temperature minimum, at which the co-ordinated locomotary movements of lizards are lost, follows an endogenous circadian rhythm in some species (Spellerberg and Hoffmann, 1972).

Animals that have evolved in situations where they are likely to be subject to environmental changes during their lifetime will have various means of adaptation, the most important of which are acclimatization and learning. McFarland and Houston (1981) discuss this topic in terms of optimality theory. Their main conclusion is that adaptation by acclimatization of learning will be beneficial in terms of fitness, but it also incurs some cost. For the change to be worthwhile, the benefit must ultimately outweigh the cost. The animal cannot afford to adapt to every small environmental change, and we can expect to find that the adaptation mechanisms offer some initial resistance to change.

The mechanisms by which animals make unanticipated adjustments to subtle changes in their circumstances are not well understood. The best we can do at the present time is to see how far the known mechanisms can take us. A radical change in the environment will induce such major changes in the animal that it will be difficult for us to tell what is happening. A small change may, as we have seen, be compensated for without much alteration in an animal's daily routine. It seems, therefore, that we should focus on the disturbance that is just large enough to induce the animal to change its behaviour.

The most fundamental type of change we have to consider is acclimatization to a changed environment. Acclimatization will result in fundamental changes in the animal's physiological state, which affect its motivational goals resulting in profound changes in behaviour. Sibly and McFarland (1974) explored the marginal conditions of acclimatization. They considered an animal acclimatized at state p (Figure 1.21). Any displacement d from p will act as an input to the acclimatizing systems, but because of the long delays that are typical of acclimatization processes, small displacements about p will have no effect, provided their time average is zero. In situations in which an appropriate behavioural response is not possible, the maintenance of the initial acclimatized state depends solely on (quick-acting) regulatory mechanisms. This means that displacements d must be within the regulatory space R, shown in Figure 1.21 (note that these spaces are parametrized by velocity vectors). Where relevant behaviour is possible the displacements that do not lead to acclimatization may occur within a space that is larger than the regulatory space, called the displacement space D. The extra contribution of behavioural mechanisms is indicated by the space B, which can take up any position between D and R. In other words, the animal's physiological state will be locally stable provided that the regulatory and behavioural mechanisms can cope with the environmentally imposed disturbance. Beyond this point gradual acclimatization will start to occur. Once this happens, the animal's whole physiological balance

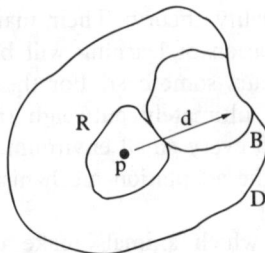

Fig. 1.21 Displacement space D, for an animal acclimatized at p. When the drift d remains within the D space no acclimatization occurs. The extent to which regulatory mechanisms contribute to this situation is indicated by the regulatory space R. The extra contribution of behavioural mechanisms is indicated by the space B, which can take up any position between D and R (from Sibly and McFarland, 1974).

will change, its motivational situation, and even its behavioural repertoire will be fundamentally altered.

The problem of motivational priorities is fitting together, both in terms of function and mechanism, all the various components of a complex system. These include scaling the short-term priorities in such a way that the animal performs the most beneficial of its possible activities at any particular time: taking account of the cost of changing between activities: arranging matters so that, in the long term, jobs are done according to their relative importance: adjusting daily routines to make the most of opportunities, but at the same time, allow sufficient flexibility to cope with environmental changes. How all these considerations are fitted together in the design of an animal remains a problem. I have tried to suggest ways of describing the components of the problem that will make this task easier.

CHAPTER TWO

Suffering in animals

The problem of animal suffering is largely a conceptual one. Human suffering is a subjective private experience. There are problems of recognizing suffering in other people, let alone other species. For more than a century scientists and philosophers have wrestled with this problem, as can be seen from Boakes' (1984) excellent historical treatment, and from Dawkins' (1980) admirable account of current attitudes to animal suffering. Basically, there are two types of approach to the problem, the common-sense approach which asks us to recognize suffering in animals just as we normally recognize it in other people, and the logical-scientific approach, sometimes called behaviourism, which usually comes to the conclusion that it is impossible to determine whether animals suffer, because it is impossible to access their private experiences. Within ethology there has been a shift in opinion, from the classical view, epitomized by Tinbergen (1951), which tried to be objective and therefore avoided the subject, to the heuristic view pioneered by Griffin (1976, 1981), which maintains that self-awareness occurs in all humans, and is likely to occur in some animals.

So far there has been little systematic attempt to link the problem of animal suffering with current ethological theory. A worthwhile theory ought to have something to say about every aspect of animal behaviour, and eventually we will have to come to one of three conclusions: (i) that the concept of suffering has no part in a scientific ethology; (ii) that there are gaps in ethological theory, within which a concept of suffering should have a place, and (iii) that the concept of suffering does indeed have a place in ethology, even though there remains much empirical ground to be covered. This paper is intended to provide a preliminary exploration of the role of suffering in ethological theory.

Attitudes to Suffering

Let us start by considering some generally recognized characteristics of suffering:

(a) Suffering is a state which results from a wide variety of circumstances.

(b) The animal will try to reduce the state of suffering.

(c) It may do this by means of species-specific avoidance behaviour, or by learned behaviour.

(d) It may, or may not, communicate its suffering to conspecifics.

(e) It may, or may not, be aware of its suffering, in the sense that people say they are.

These general characteristics give us a rough idea of what we mean by suffering in animals. The form of a more precise definition will depend largely upon our attitude to suffering, the type of approach, or stance, that we wish to take. Within the scientific world there are three main attitudes, corresponding to the three major disciplines that are relevant to the problem. Thus if we ask how we are to know whether an animal is suffering, we will find three general types of answer: (i) we might find direct physiological evidence; (ii) we might feel philosophically obliged to assume suffering, or (iii) we might find behavioural evidence. These correspond to the most typical types of stance found in the literature on animal suffering.

The physiological stance

We can measure physiological indices of autonomic activity and stress in animals, but do such measurements permit us to conclude that the animal is suffering? We can do this only by analogy with humans. In other words, if humans have a particular physiological response when they report that they are suffering, and if we detect the same response in an animal, then we might conclude that the animal is also suffering.

To do thus convincingly we have to overcome a number of objections:

(a) The human verbal reports may not be reliable.

(b) The animal may show the physiological response, but its brain may not be capable of experiencing suffering.

35

(c) The animal may suffer without showing any physiological signs. Autonomic arousal occurs in a variety of situations, some of which may not involve suffering. Moreover autonomic activity does not show obvious differences in different situations (Dawkins, 1980).

(d) The apparent suffering may not influence the animal's behaviour in any way that we can detect. I think that we must insist on some demonstration that the postulated suffering affects behaviour. In other words, if we can account for avoidance, communication and other behaviour without invoking suffering, it would seem that the concept is superfluous.

Philosophical views

There are many different philosophical views on animal suffering, emotion, and related topics. I do not propose to review these, but comment on a few so as to clarify my own position. A common view is that although suffering may not be causally implicated in behaviour, it exists as an epiphenomenon, or side effect.

I see three problems here: (i) if suffering does not influence behaviour (and/or physiology), then it is unobservable, and not open to scientific inquiry; (ii) if suffering does not influence behaviour then is not necessary for the explanation of behaviour; (iii) if suffering does not influence behaviour, then it is difficult to see how it could have evolved.

The concept of suffering has value for ethologists only if it can be shown to be necessarily instrumental in the causation of behaviour. This means that, in testing any given hypothesis, we must ask ourselves whether we can satisfactorily explain the behaviour without invoking suffering. Unless we do this we are in danger of creating a metaphysical cul de sac.

Another philosophical stance is that animal suffering is somehow connected with self-awareness or consciousness as manifest in humans (Griffin, 1976, 1981). It seems to me that it is presumptuous for us to assume that because our suffering involves self-awareness, this should also be true of other species.

This stance implies that only animals capable of self-awareness can suffer. The main argument in favour of this view is that we cannot imagine suffering without self-awareness. This is a strange argument for an ethologist. In discussing any other type of behaviour the ethologist would not think of using his or her own feelings as part of

36

a scientific argument. Ethology, as an objective science (Tinbergen, 1951) is supposed to shun anthropomorphism. Why should an exception be made in this case? We do not have to imagine subjectively the suffering (if any) being experienced by the animal. We should recognize that, like many aspects of animal perception, it may be very different from our own.

The behavioural stance

We are looking for behavioural evidence of a particular type of internal state, and of its mental counterpart. We do not have, nor can we have, direct behavioural evidence as to an animal's internal state. The state in question is an hypothetical construct. We have good general grounds for pointing to changes in state as part of our explanation of changes in behaviour (Hinde, 1985; McFarland and Houston, 1981), but to make claims about a particular aspect of the animal's state we have, inevitably, to enter into a certain amount of model-building.

To claim behavioural evidence of an hypothetical construct it is necessary to show that the behaviour in question could not be explained without invoking the construct. In other words, to have behavioural evidence of suffering it would be necessary to come to the conclusion that the behaviour in question could not be explained without invoking a concept of suffering.

As an example, let us consider the question 'Do animals communicate their suffering to each other'? We can often tell when an animal is in a state of fear, sickness or frustration, so it is not surprising that other members of the same species can also tell. In many species these states are accompanied by ritualized forms of communication, such as facial expression, vocalization, or pheromones.

Charles Darwin (1872) claimed that the so-called emotional states are accompanied by specific forms of communication, and recent workers have tried to establish a link between communication and awareness. One suggestion is that, in responding to the emotional communication of a conspecific, an animal is better equipped to understand the message if it can identify with the emotional state of the other (Humphrey, 1978; 1979). This awareness of the feelings of others, which enables the recipient to identify with the other animal, is supposed to require some form of consciousness, and presumably includes suffering.

Apart from the question of whether awareness is a relevant issue (see above), there is also a difficulty with another aspect of this argument. It may be that empathy with the feelings of others makes a difference to our behaviour, but how? It might, but is there any compelling reason why it should?

The problem is that for each and every specific situation for which an explanation involving suffering is offered, there is always an alternative procedural explanation. For example, where one animal comforts another in distress, we might want to conclude that the comforter shows empathy with the sufferer. On the other hand we might want to conclude that the comforter responds to the distress signals by following certain procedural rules. We can imagine that there could be sufficient evolutionary pressure for such rules to become established in a species with a rich social and political life. As we see below, if we can account for behaviour in terms of procedural rules, we do not require a concept of suffering.

Another obvious possibility is that suffering is involved in the causation of avoidance behaviour. Animals may naturally avoid situations that they fear, and they may learn to avoid situations as a result of experiencing pain, sickness, or frustration. Can we explain these aspects of behaviour in both causal and functional terms without invoking suffering?

Certainly, we can account for avoidance of particular stimuli without invoking suffering. We can simply say that the animal has acquired a tendency to avoid the stimulus, either in phylogeny or by learning. Explanations in terms of species-specific defence reactions (e.g. Bolles, 1970), or in terms of conditioning (Mackintosh, 1983), do not depend upon any concept of suffering.

Similarly, when we think in evolutionary terms, we find it difficult to think of a good reason why animals should be designed to suffer from particular stimuli, rather than simply avoiding them, or learning to avoid them, in an automaton-like manner.

However, in the case of avoidance of poisons, or of vitamin-deficient foods, the situation is not so clear cut. Most workers in this field agree that, with the exception of certain specific factors such as sodium, the characteristic of physiological state that is important in this type of problem is a general state of 'sickness', or 'well-being', rather than specific information concerning the nature of the dietary deficiency or toxicosis (Rozin and Kalat, 1971; Rozin, 1976a). Thus explanations of this type of avoidance depend upon a general concept of sickness, which may not be very different from the concept of suffering that we are seeking to investigate.

Procedural and Non-Procedural Explanations

So far, we have discussed suffering in situations which are always open to a particular objection. The objection is that suffering is not a concept that is necessary for explaining the behaviour, because in each particular case the animal has a ready and automatic response, or is able to quickly learn an appropriate response. The animal responds on the basis of procedural rules which are either innate or acquired by simple conditioning.

A possible exception, mentioned above, is food aversion learning. Here, the animal learns to avoid a particular food by association with a general state of sickness. If it could be shown that the animal would learn to do anything (not only avoid the food) to avoid or reduce sickness, then we might have grounds for saying that a set of procedural rules is not adequate to account for the behaviour.

In the procedural case the animal responds to the situation, be it a simple stimulus or a complex state of affairs, in an automatic manner without monitoring the consequences. The behaviour results in consequences that impinge upon the animal, but the animal does not guide its behaviour on the basis of the consequences. The behaviour is designed to occur in a particular type of situation, so the consequences will generally be the same on different occasions. In other words the consequences are a design feature of the procedural rules, as illustrated in Figure 2.1. For example, a moth may have a rule which says 'if disturbed when resting then spread wings'. A consequence of spreading the wings is that eye spots are revealed. A consequence of this is that the disturber, usually a small bird, is frightened away. It may be that, on a particular occasion, the disturber is not a small bird. Nevertheless, the procedural rule is designed on the assumption that moths are generally disturbed by small birds, and when they are disturbed by other things it does not really matter, because few other animals eat moths that are resting on vegetation. An implicit assumption in this type of procedural explanation is that the circumstances in which the behaviour occurs have arisen sufficiently often in the history of the species for an appropriate

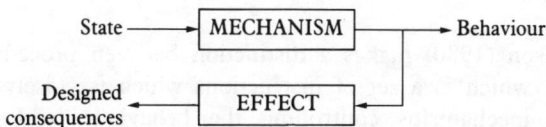

Fig. 2.1 Diagram of a procedural mechanism in which the consequences of behaviour are evolutionarily predicted.

response to have been designed by natural selection, either as an innate response, or as a readily learned response.

The essence of the non-procedural case is that the animal has no automatic response to the situation, and must guide its behaviour on the basis of the consequences, as illustrated in Figure 2.2. To do this it must have a representation of the desired consequences, and it must compare this with the actual consequences gained from the behaviour. As a result of this comparison the animal guides its behaviour. Although this may appear to be very similar to a simple servosystem, these are some important differences.

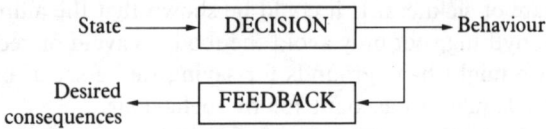

Fig. 2.2 Diagram of a non-procedural mechanism in which the consequences of behaviour are under the animal's control.

In a simple servosystem, such as that illustrated in Figure 2.3, the behaviour is actuated by the error-signal, which is the output of a comparator that is a simple summing device. The servo-system can itself be regarded as a procedural rule, which is activated, automaton-like, under certain circumstances. Many such cases have been documented in animal behaviour (McFarland, 1971; Toates, 1980). In a non-procedural feedback device, comparison of desired and actual consequences does not lead to an automatic response. This difference from simple procedural feedback has to do with the way the relevant information, or knowledge, is represented in the animal.

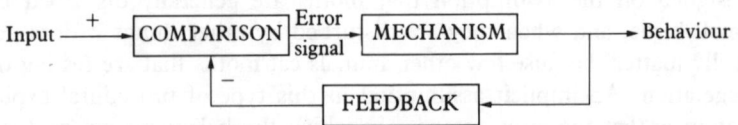

Fig. 2.3 Diagram of a simple servomechanism in which the behaviour-control mechanism is actuated by an error-signal.

Dickinson (1980) makes a distinction between procedural representation, which is a set of instructions which is closely integrated with the mechanisms controlling the behaviour, and declarative representation, which describes a relationship between events in the animal's world, but does not commit the animal to using the information in any particular way. Dickinson (1980) illustrates this distinction by reference to work on animal learning. Holland (1977) exposed rats to a tone-food relationship by presenting an 8-sec tone and then presenting food pellets. The rats developed a tendency to approach the food receptacle during the tone, suggesting that a procedure for approaching the food receptacle is established, so that the learned information is stored in a form that is directly related to its use. The alternative possibility is that the rat learns that 'tone causes food', thus establishing a declarative representation. Because a declarative representation is passive, and does not control the animal's behaviour, the rat's tendency to approach the food during the tone would have to be accounted for in some other way. The procedural hypothesis thus provides a more parsimonious explanation of the rat's behaviour. Suppose, however, that once the tone-food association is established, the rats are exposed to a food-illness relationship to the point where they refuse to eat food when it is presented. The rat will then have formed two different associations, tone-food and food-illness. A procedural theory would imply that the rat could not integrate the two procedures, which are quite separate recipes for action. A declarative theory, on the other hand, provides a basis for integration because food occurs in both (tone causes food, food causes illness), and because the representation does not specify any particular action. Holland and Straub (1979) showed that rats can integrate information from associations learned at different times. Rats exposed to food-illness following tone-food were disinclined to approach the food when the tone was presented again.

An animal may respond in a procedural manner to a stimulus that it has not encountered before, or which has not previously been encountered in the history of the species. By stimulus generalization the animal may interpret the novel stimulus in terms of something with which it is familiar, and then it may invoke a procedural rule from its normal repertoire. However, it does seem to me plausible to argue that there are certain circumstances in which the animal is unable to perform its natural behaviour, that could not have arisen in evolution, and about which the individual animal cannot generalize on the basis of its past experience.

Procedural explanations are adequate in situations where the behaviour could plausibly be due to the application of a simple rule by the animal. The rule may involve a simple stimulus response relationship, it may incorporate a purpose-designed servosystem, it may involve stimulus generalization, classical conditioning, etc. It may even involve contingent rules.

41

Procedural explanations are not adequate in cases where:

(a) There is reason to believe that the relevant rule could not have become incorporated into the animal's repertoire

(b) The animal is able to combine previously unrelated aspects of its behavioural repertoire

(c) The behaviour involves an innovation, or provides a new solution to a problem.

Suffering as Motivational State

If suffering is a state of the animal that can influence behaviour, it must be representable in a state-space. In other words, the animal's physiological state which gives rise to the brain-state that we may or may not wish to call suffering, can be portrayed in terms of the state-space described by McFarland and Houston (1981). Figure 2.4 shows a simple version of this state-space. The obvious first suggestion is that suffering is associated with the lethal boundary. As the state approaches the lethal boundary more and more desperate physiological and behavioural mechanisms come into play, as illustrated in Figure 2.5. Many of these mechanisms will be very stereotyped and specific, such as panting in birds. If these automatic restorative mech-

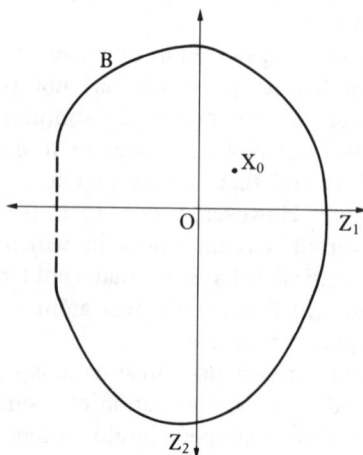

Fig. 2.4 Diagram of a (two-dimensional) physiological state-space. O = origin, Z_1 = and Z_2 are axes, B is the boundary to the space and X is the state at a particular point in time.

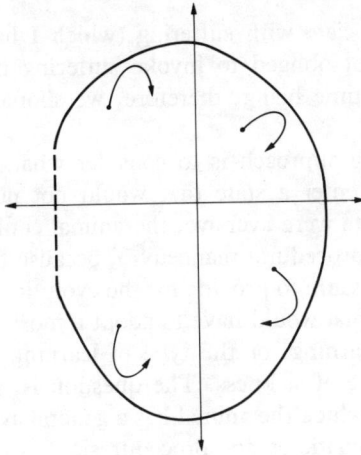

Fig. 2.5 Trajectories in the state-space are repelled from the lethal boundary by physiological emergency mechanisms.

anisms break down, or fail to stop the drift in state, then the animal will die.

Although the animal may be suffering, we cannot say that suffering is necessary to explain the behaviour, because we are already explaining the behaviour in terms of automatic procedures. So the region of the space near to the lethal boundary, illustrated in Figure 2.6, is a region in which the animal behaves in a manner that we

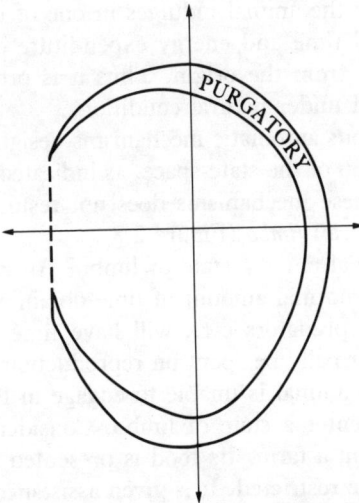

Fig. 2.6 Region of the state-space, called PURGATORY, in which the animal is under physiological strain.

43

would normally associate with suffering (which I have called *purgatory*), but we are not obliged to invoke suffering in explaining the behaviour. For the time being, therefore, we should look elsewhere in the state-space.

A more promising approach is to consider what would happen if the animal were to enter a state that would not normally occur in nature. If such a state were aversive, the animal could not change its state by means of a procedural manoeuvre, because there would have been no selective pressure to provide for the evolution of the necessary procedures. The animal would have to adopt a more general strategy, such as trial-error-learning, or the type of learning that is normally involved in avoidance of sickness. The question is, is there a region of the state-space to which the animal has a general aversion, but from which it has no automatic escape procedures?

A possible candidate fulfilling these criteria is the region of the state-space near to the origin. In this region the animal has no physiological imbalances. It does not need to do themoregulatory, feeding, drinking or any other homeostatic type of behaviour. It therefore has more time for other types of behaviour. We would expect, from an evolutionary point of view, that the animal would be designed to be motivated towards reproduction, hibernation or migration, when its state entered this region. In other words, when food is plentiful the time and energy is available for those activities that more directly promote reproductive success, or ensure survival in anticipated hard conditions. As soon as the animal indulges in one of these activities, then the consequential time and energy expenditure ensure that the trajectory moves away from the origin. Thus it is probable that the origin is never reached under natural conditions.

Thus there are various automatic mechanisms designed to keep the animal out of this region of the state-space, as indicated in Figure 2.7. However, failure of these mechanisms does not result in death, but in a state which I will call *limbo* (Figure 2.8).

How might an animal enter a state of limbo? An animal that has to spend less than the normal amount of time obtaining food, water, shelter, freedom from predators etc., will have time and energy to spare. These would normally be spent on reproduction, migration, or hibernation, but if the animal is unable to engage in these activities, then it will probably enter a state of limbo. Consider an animal in captivity, in a zoo or on a farm. Its food is presented on a plate. Its social circumstances are restricted. It is given assistance in combating disease. Physiologically, it is likely to be approaching limbo.

Normally, the animal's reproductive aspirations, and the hormonal

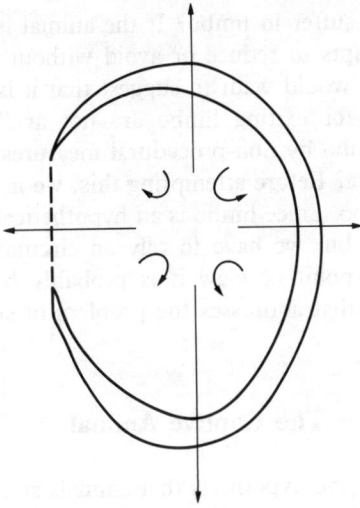

Fig. 2.7 Trajectories in the state-space turn back from the origin by physiological mechanisms designed to take advantage of the favourable situation.

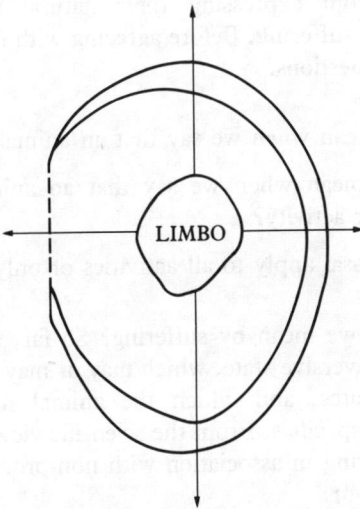

Fig. 2.8 Region of the state-space, called LIMBO, which is not usually penetrated by the trajectory under natural conditions.

and energetic changes that accompany them, would prevent the state from entering limbo, but this does not occur in the captive animal, which cannot migrate, engage in much reproductive politics, experience normal parental behaviour, etc.

Does the animal suffer in limbo? If the animal is in a state which it persistently attempts to reduce or avoid without the help of automatic procedures, I would want to suggest that it is suffering. If the normal procedures for exiting limbo are not available, the animal could only avoid limbo by non-procedural measures. Can we demonstrate such avoidance? Before attempting this, we have to know when the animal is in limbo. Since limbo is an hypothetical state, we cannot monitor it directly, but we have to rely on circumstantial evidence. From the practical point of view it is probably better to set up a working hypothesis that addresses the problem of suffering directly.

The Captive Animal

Suppose we advance the hypothesis that animals suffer when they are unable to do what they are designed to do. This is not a new idea. It was raised by Thorpe (1965) in his testimony to the Brambell (1965) Committee. Thorpe argued that keeping animals in ways that prevented them from expressing their natural instinctive urges undoubtedly caused suffering. Before agreeing with this view we must face three crucial questions:

(a) What do we mean when we say that an animal will suffer?

(b) What do we mean when we say that an animal is unable to perform a particular activity?

(c) Does the proposal apply to all activities or only to some?

Firstly, what do we mean by suffering? So far, all we can say is that suffering is an aversive state, which may or may not have distinct physiological correlates, and which the animal may or may not communicate to conspecifics. From the scientific viewpoint, we would expect to find suffering in association with non-procedural aspects of the animal's behaviour.

Secondly, what is meant by unable to perform an activity? An animal will suffer, according to our hypothesis, if it is unable to perform an activity, and would have performed it otherwise. An animal may not perform a particular activity because: (i) some other activity has priority; (ii) the cost of changing is too high; (iii) the relevant stimuli are not present; (iv) it is physically prevented.

Let us consider some brief examples:

(a) (Some other activity has priority). Food is not readily available, so the animal has to spend a lot of time foraging, leaving less time for other activities, and perhaps no time for some. Hunger maintains priority over the other motivations, and some activities are not performed because of lack of sufficient motivation.

(b) (The cost of changing is too high). A foraging animal becomes thirsty, but water is a long way away and would require considerable effort to reach. The cost of changing from feeding to drinking, in terms of time, energy, and opportunity loss, may be so high that drinking is postponed.

(c) (The relevant stimuli are not present). An animal's internal state may be appropriate for a particular activity, but the relevant external stimuli are lacking. Since the strength of internal stimulation is usually an important contributor to the tendency to perform the activity, there will usually be insufficient motivation for this activity to compete with others. However, it may be that the internal factors are so strong, or competing tendencies so weak, that the animal is simply unable to perform an activity that it otherwise would perform.

(d) (Physical prevention). All the requisite causal factors may be present, and the animal may be strongly motivated, but unable to perform the activity because it is physically prevented from doing so. In the short run, we would expect this situation to lead to frustration, but in the long run we would expect the animal to acclimatize to the situation, although still maintaining its unfulfilled motivation to perform the activity. In other words, the animal has certain physiological acclimatization mechanisms (especially with respect to such variables as ambient temperature, level of nutrition etc.) which enable it to adjust certain aspects of its physiological state (such as reproductive state), so that its motivational repertoire is changed (McFarland, 1985).

In the first two cases, the animal has a certain freedom of choice. "It could perform the activity if it wanted to." A better way of putting this point is that evolution will have designed the animal to cope psychologically with these contingencies (provided, of course, that they are similar to contingencies that were prevalent in the species' past). We cannot say that the animal is unable to perform the activity, and we would not expect it to suffer. In other words, we would not expect an animal to suffer as a result of deciding not to do something because it was too hungry, or because the cost of changing was too high.

The third case is more difficult. If the animal's motivation to perform a particular activity is below that of other activities, because of lack of relevant external stimuli, then the animal will be busy doing the other activities and may not have performed the activity even if the stimuli had been present. We would not expect it to suffer simply as a result of lack of relevant stimulation, but only if the animal would have performed it otherwise. Even then, we might want to say that an animal was designed to cope with the lack of stimulation. An ambush predator awaiting some prey, presumably does not suffer, whereas a grazing animal deprived of grass might suffer.

An animal that is strongly motivated to perform an activity, but is physically prevented from doing so, will generally become frustrated. This is a particular motivational state, characterized by certain physiological responses, and by a number of behavioural signs of 'agitation', such as displacement activity, aggressive behaviour etc (see McFarland, 1981, 1985). This may result in the animal trying harder initially, but then breaking away and performing other activities. If the frustrating situation persists, then the animal will generally learn that it is a no-go situation and come to ignore it. However, in some situations the animal's strong motivation may persist, and we should consider whether such circumstances involve suffering.

In short, my proposal is as follows: an animal may suffer if it is unable to perform an activity that it is motivated to perform. In cases that are likely to be commonplace in nature, we might expect the animal not to suffer because it has been designed to cope with this type of situation. Even if the animal does suffer in such circumstances, we cannot counter the objection that it may follow a simple procedural rule, and that the concept of suffering is superfluous as an explanatory concept.

However, there are two types of situation in which it is more difficult to sustain the procedural argument. Firstly, the animal may not be able to perform an activity because the relevant stimuli are, contrary to the natural situation, not present. Secondly, the animal may not be able to perform an activity because it is physically prevented in unnatural circumstances.

How might an animal come to avoid situations that do not arise in nature? First, it enters a particular state as a result of some general consequence of the situation. This particular state must then be instrumental in the avoidance behaviour. A similar situation arises in food-aversion learning. How does an animal come to avoid the wide variety of substances that might make it ill if eaten? It does not have a specific avoidance mechanism for each one. Instead it enters a generalized state called *sickness* which is instrumental in the subsequent avoidance behaviour. I am suggesting a somewhat analogous state, called *suffering*, which is similarly instrumental in avoidance behaviour.

The difference between sickness and suffering is that the former is easy to detect physiologically, whereas the latter is not. Imagine that we could not detect sickness physiologically. How would we know that it was occurring? We could observe that the animal avoided some feeding situations, but not others. We could discover experimentally that certain manipulations caused the animal to avoid a previously favoured feeding situation. Thus injection of lithium chloride after a meal makes the animal subsequently avoid the type of food eaten in that meal. Even if we did not know that lithium chloride made the animal sick, we might discover that many different types of manipulation caused food aversion. If all these manipulations were natural ones, it might be possible to argue that the animal had been designed to avoid, or to learn to avoid, each of those specific situations. However, it would be difficult to argue this way if the manipulations could not have been encountered during the course of evolution. It would be more plausible to argue that the manipulations all contributed to a general state, called *sickness*, or *suffering*.

To summarize this part of the argument, an animal may, or may not suffer, if rendered unable to perform a particular activity. To avoid the procedural type of objection, we should devise means of preventing the animal from performing the activity that would not have occurred during the evolution of the behaviour.

Finally, does the proposal apply to all activities or only to some? The hypothesis that animals suffer when unable to do what they are designed to do may not apply to all activities. It has been said that animals would not suffer if prevented from engaging in antipredator behaviour (Dawkins, 1980; Hughes and Duncan, 1981; Perry, 1978), but I am not sure that this is a valid assumption. I could well imagine an animal feeling uneasy if it could not carry out its daily ration of anti-predator activity. The point is that we cannot judge this issue *a priori*. We need ways of finding out which activities are important to the animal, and which are not.

Traditionally, deprivation experiments have been the standard method of tackling this question. The basic idea is that an animal deprived of an activity that is important will attempt to catch up as illustrated in Figure 2.9. Thus much research has been done to find out the extent to which we catch up on missed sleep.

The basic problem with this approach is that an apparent

DEPRIVATION

PROBE

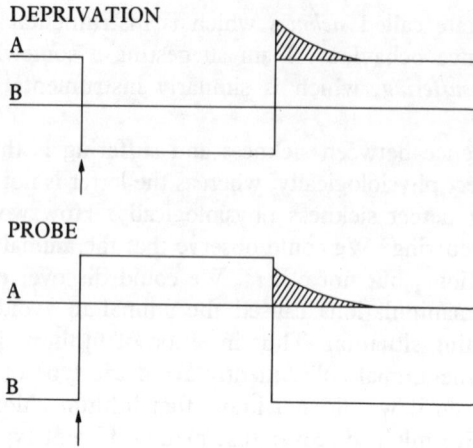

Fig. 2.9 Schematic representation of deprivation and probe experiments. In a deprivation experiment (*above*) the top-priority motivational tendency (A) is reduced by experimental manipulation. In a probe experiment (*below*) a low-priority motivational tendency (B) is increased by experimental manipulation. The arrows indicate the onset of experimental manipulation. Shaded areas represent possible damming-up or rebound effects.

damming-up effect can result from either an accumulation effect, or from a rebound effect. In the first case, the motivation to perform the deprived activity increases during the deprivation period, while in the second case it decreases due to habituation. The rebound is due to dishabituation. It is often assumed that an accumulation of drive or behaviour potential is responsible, but some cases are almost certainly due to a rebound effect. For example, when a child is given a new toy it plays with it a lot initially, and then it settles down to a baseline level. Suppose the toy disappears and the child does not notice. When the toy is reintroduced, the level of play will immediately rise above the baseline. This is almost certainly due to a rebound effect. The child plays at the baseline when he is familiar with, and habituated to, the toy. During the period of deprivation there is dishabituation, so that when the toy reappears it is something of a novelty.

I cannot help thinking that a similar explanation might account for the increase in dust bathing by fowl following periods of dust deprivation (Vestergaard, 1980, 1982), and perhaps for other cases where habituation to the stimulus is a plausible possibility.

The basic problem is that we do not know what is going on during the deprivation period. The assumption that there is a build up of tendency to perform the deprived behaviour may not be valid. Indeed, in the only case that I know of where a measure was made of the tendency during deprivation, the result was the opposite of the expected (Heiligenberg and Kramer, 1972). There are also problems in measuring recovery, and with possible disturbing effects of the deprivation procedure. For example, even in the case of food deprivation, recovery may not be a reliable measure (McFarland, 1965), because the animal may eat less in recovery from prolonged deprivation than after a shorter deprivation. The deprivation procedure may cause stress, often the case in sleep research, or, where deprivation is by removal of key stimuli, it may result in dishabituation. Both these confound the issue.

The problem is partly that, at the onset of deprivation, the animal is given no choice. An alternative procedure is to use a probe instead of forced deprivation. A probe is a stimulus that is presented to an animal, offering it an alternative activity, but in no way preventing the animal from continuing with its current behaviour (see Figure 2.9).

For example, in his investigations of fanning in sticklebacks, Sevenster (1961) compared the effects of introducing a female in a glass tube for 5 min, or a rival male in a glass tube, or physically preventing fanning by covering the nest with glass. An example of his results is illustrated in Figure 2.10. Damming-up effects were obtained in all three cases. These results show that inducing the

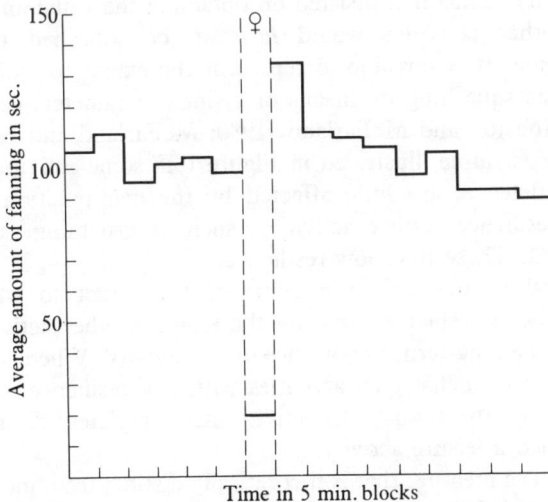

Fig. 2.10 An example of a probe experiment. Interrupution of male stickleback fanning behaviour by presentation of a female in a glass tube for 5 minutes (during which the male courts the female). Notice the increase in fanning after the interruption (after Sevenster, 1961).

animal to take up an alternative activity can produce an apparent damming-up effect.

Another approach is to try to measure the overall importance of the activity to the animal. We can expect that activities which promote the survival of the individual to take precedence over other activities, such as territorial behaviour, whose contribution to fitness is less immediate. Each activity has some value in terms of fitness and animals must somehow allocate priorities among activities. One approach to this problem is to devise a measure of the cost to an animal of abstaining from each activity in its natural repertoire. If an animal did no feeding, for instance, the cost might be high, but the cost of abstaining from grooming would probably be low. An animal with high motivation both to feed and to groom, but insufficient time to do both, would sacrifice less fitness if it devoted its time to feeding at the expense of grooming.

Figure 1.16 illustrated the way in which an (hypothetical) animal might spend its time in a particular environment. If the environment is much the same from day to day then the animal will settle down to a daily routine (McFarland, 1985) that was also much the same from day to day. If there were a change in the environment such that it now takes much longer to obtain the same amount of food (Figure 1.17), then the animal would have to adjust by altering the time spent at each activity. Thus if it insisted on obtaining the usual amount of food, the other activities would have to be squashed into the remaining time. It is possible to represent the extent to which each activity resists squashing by means of a single parameter, called the resilience (Houston and McFarland, 1980; McFarland and Houston, 1981). In the example illustrated in Figure 1.17 some activities, such as territory defence, are little affected by the manipulation. These have high resilience. Other activities, such as grooming, are very much affected. These have low resilience.

Behavioural resilience is a measure of the extent to which an activity can be squashed in time by the animal's other activities. It thus reflects the long-term importance of the activity. When time acts as a constraint on behaviour, activities with low resilience will tend to be ignored by the animal. An activity that completely disappeared might be called a leisure activity.

Resilience is a measure that is theoretically distinct from motivation (McFarland and Houston, 1981), but it is sometimes difficult to isolate its effects (McFarland, 1985). Few attempts have been made to measure resilience directly, because these require prolonged observation and manipulation of the animal's behaviour throughout the day. However, resilience can be measured indirectly by means of demand functions.

Demand functions are used by economists to express the relationship between the price and the consumption of a commodity, as illustrated in Figure 2.11. Demand is said to be elastic when consumption falls as the price rises, and inelastic if consumption is little affected by price changes. Analogous phenomena occur in animal behaviour. Numerous studies (Lea *et al.*, 1987) show that the amount of activity shown (the consumption) varies with the required energy expenditure (the price) in a manner that is easily recognized as a demand function. An example is illustrated in Figure 2.12. This parallel between the demand phenomena of animals and people is a topic of considerable current interest.

The elasticity of demand functions gives an indication of the relative importance of the activities (or commodities) on which the animal (or person) spends its energy (or money). There is a close relationship between elasticity of demand and resilience (Houston and McFarland, 1980). If activity A has lower resilience than activity B, then A will show a more elastic demand function than B (McFarland and Houston, 1981). Thus demand functions can be used as an indirect measure of resilience. The main problem with demand functions is that elasticity increases with the availability of substitutes.

For example, Lea and Roper (1977) found that demand for food was inelastic in rats required to work to obtain food rewards. As the work required increased, the rats continued to work for about the same amount of food, thus showing inelastic demand for food.

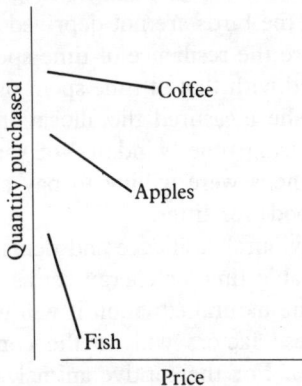

Fig. 2.11 Examples of economic demand curves. Demand for coffee is inelastic, demand for fish is elastic (both axes have log scales and arbitrary origins) (from McFarland, 1985).

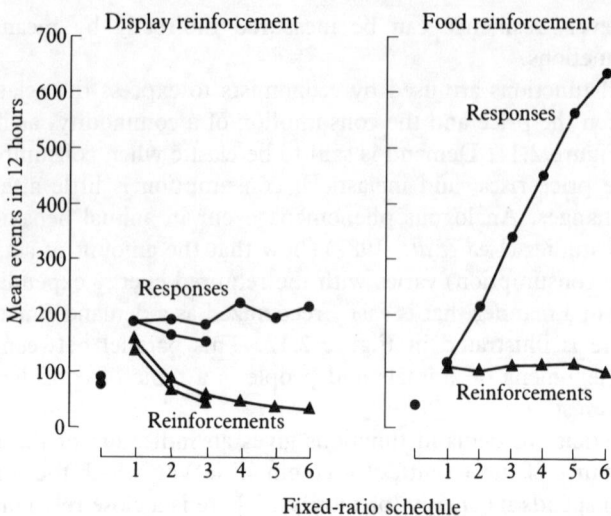

Fig. 2.12 Mean response rates of groups of Siamese fighting fish (*Betta splendens*) swimming through a tunnel for food reward (*right*), or for the opportunity to display to their mirror image (*left*). The abscissa shows the number of responses required to obtain a food reward (from McFarland, 1985, after Hogan, Kleist and Hutchins, 1970).

However, they also found that the elasticity of demand for food pellets increased when sucrose was available as a substitute. Thus, as a measure of the importance of the activity to the animal, the demand measure is contaminated by the substitution effect. Unless the availability of substitutes is known, the demand function is difficult to interpret. Dawkins (1983) showed that, although hens prefer litter to wire floors, their preference is not strong enough to outweigh the attraction of food unless the birds are not deprived of food at all. She also attempted to measure the resilience of time spent on litter in the absence of food compared with that of time spent on wire in the presence of food. (Strictly, she measured the allocation of time between these two possibilities during time 'windows' of varying length). She found no evidence that hens were willing to pay a cost (in terms of time spent away from food) for litter.

The concepts of behavioural resilience and demand are relevant to situations in which available time or energy act as constraints on the animal's behaviour. In the natural situation it will usually be the case that one or both of these factors will be the constraint that bites (McFarland, 1981, 1987). For the captive animal, however, there is no shortage of time or energy. Such animals receive sufficient food from their captors without having to spend time foraging. Their behaviour is usually curtailed in other respects, so they have time and energy to spare. In terms of ethological theory, the situation where the animal has too little time to do all that it might otherwise do is relatively straightforward compared with the inverse situation. If we take the hypothetical case illustrated in Figures 1.16 and 1.17 at face value, we can extrapolate to the condition where the animal obtains all its food in half an hour instead of the normal 8 hours. When we do this (Figure 2.13) we see that time spent in sleep increases from 7.5 to 10.5 hours, territorial defense from 4 to 5 hours, drinking from 2 to 3 hours, and grooming from 2.5 to 5 hours. However, a linear extrapolation is not really justified, because in theory (see McFarland and Houston, 1981) the extent of over-indulgence in certain activities will depend upon the associated costs. In other words, for some activities, such as drinking, it does not matter much if the animal does more that is usual, but for other activities over-indulgence may be deleterious. This is a little explored area and one which needs to be tackled if the problems confronting animals in captivity are to be addressed.

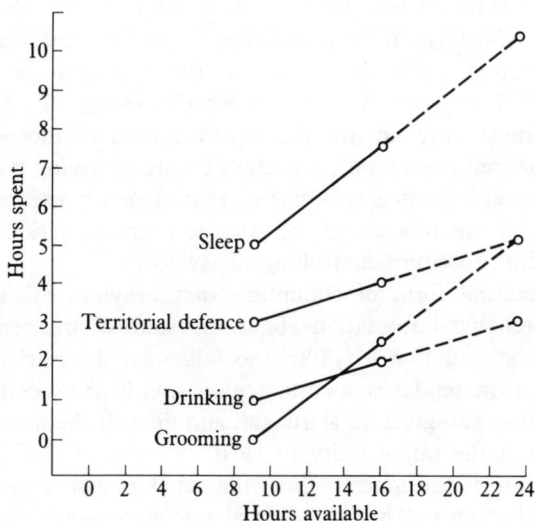

Fig. 2.13 Hours spent at each activity once the hours spent feeding are accounted for. The solid lines join the data points given in Figs 1.6 and 1.7. The dotted line is an extrapolation to the case where the animal obtains all its food in $\frac{1}{2}$ hour per day.

Suffering and Communication

As we have seen (p. 37) some animals appear to communicate their distress to each other, in the form of vocalizations, facial expression or pheromones. The function of such signals is not always obvious, but it is likely that the distress signals broadcast by an animal being attacked by a predator, may serve as a warning to conspecifics, as in the warning substances released by some ants and fish, or to solicit aid from members of the group.

We are not justified in concluding that distress signals necessarily indicate that the animal is suffering, because the vocalization or facial expression may occur on a purely procedural basis. Only if we had grounds for believing that distress signals had a declarative basis could we begin to think of them as evidence of animal suffering.

The distinction between procedural and declarative communication can be illustrated in anthropomorphic terms. In the procedural (knowing how) case, it is as if the animal said "I know how to get out of this mess". In the declarative (knowing that) case it is as if the animal said "I know that I am in a mess". The procedural animal may yell as part of a routine for soliciting aid or warning others, but what about the declarative animal?

Elsewhere (McFarland, 1988), I have argued that there is an evolutionary advantage in the teleological form of communication, i.e. in communication involving reference to future goals-to-be-achieved. For example, I may say "I aim to finish this essay". To be able to say this, I must have an internal representation of the goal-to-be-achieved (goal-representation), namely the state of having finished the essay. I have also argued (elsewhere) that such representations are declarative and not procedural, i.e. the goal-representation does not form part of a procedure-controlling behaviour.

The procedural form of communication provides other animals with fairly reliable information about the state of the sender, and about the behaviour that it is likely to follow. If the communication is honest then the sender is open to exploitation by conspecifics. Thus if an incubating gull gives an alarm call and flies off the nest, a neighbour can seize the opportunity to steal its eggs. If the procedural communication is dishonest, then the sender will eventually be discriminated against. Thus the animal that persistently cries 'wolf' to gain an advantage, runs the risk of having its communication devalued or discounted (see also Chapter 6). The advantage of the declarative communication (e.g. "I am suffering") is that it gives no clear indication of what behaviour is likely to follow. Thus it is difficult for rivals and enemies to take advantage of the situation, especially if there is no indication of what the actor is suffering from.

In highly social animals, the politics of suffering are likely to be complex. If the sufferer is injured it might be in its interest to receive aid from conspecifics. Whether it is in the interests of others to give aid, is another matter. If the sufferer has a disease, it might be in its interests to keep quiet so as to spread the disease to rivals. There is always a chance that it will recover and inherit a dead rival's property. This scenario might also work the other way around. Thus it might be in an animal's interest to declare that it was "very well", even if it were suffering, as has been suggested for some instances of sexual selection (Hamilton and Zuk, 1982). It is important to distinguish, however, between evolutionary strategies carried out in a procedural manner, and individual strategies with a declarative basis.

In recent years there has developed a considerable interest in the topic of deception (Mitchell and Thompson, 1986; Trivers, 1985). This follows from evolutionary arguments to the effect that communication is not simply the transmission of information, but can be expected to be designed (by natural selection) to manipulate other individuals (Dawkins and Krebs, 1978). However, it is evident that a large proportion of the deception practised by animals is entirely procedural, as in the numerous instances of mimicry (Trivers, 1985). Some of the literature seems very confused and equivocal on this point (see Chapter 6), and much of the evidence is anecdotal. For example, de Waal (1982) in his study of chimpanzee social behaviour notes that "Yeroen hurts his hand during a fight with Nikkie. . . Yeroen walks past the sitting Nikkie from a point in front of him to a point behind him and the whole time Yeroen is in Nikkie's field of vision he hobbles pitifully, but once he has passed Nikkie his behaviour changes and he walks normally again. For nearly a week Yeroen's movement is affected in this way whenever he knows Nikkie can see him." It is possible to imagine both procedural and declarative explanations of these data. How these might be distinguished remains an open question, but it is no good being prejudiced and simply jumping to one conclusion or the other. It is good that a effort is being made to tackle this type of problem (Byrne and Whiten, 1985, 1988; Whiten and Byrne, 1988).

In relation to animal suffering, these developments are important because they open the way for a possible diagnosis. Let us imagine that it could be established that Yeroen's attempts to deceive Nikkie were not purely procedural routines. By limping whenever he thinks Nikkie can see him, Yeroen is effectively saying "See how I am suffering". There are a number of possibilities here: (i) Yeroen could really be suffering pain, but chooses to hide this when out of Nikkie's presence; (ii) Yeroen's original suffering could have abated, but he chooses to display his memory of it when in Nikkie's presence;

(iii) Yeroen may not have suffered initially, but chooses to pretend that he did, and that he is suffering still. In all these cases, however, Yeroen must have some (mental) representation of the suffering (this need not be so in the procedural case). For Yeroen to be able to say "See how I am suffering," he must have a declarative representation of suffering that is not tied to any particular procedure. What Yeroen decides to do with this representation (i.e. advertise it, or hide it) is another matter. The important task, therefore, is to find ways of deciding what counts as evidence for declarative representation. This topic is discussed further in Chapter 6.

Conclusions

The problem of animal suffering is that it is open to doubt as a necessary scientific concept. It is unfortunate, in view of its probable effect on animals, that it should be open to such prolonged doubt. There are those who say that we should give the animals the benefit of the doubt and proceed as though we knew that animals suffer in the way we (think we) do. This may be morally satisfactory as a short-term approach, but it is open to question on many counts. Do all animals suffer as we do? Do all suffer in the same way? How can we tell when they are suffering and when shamming? In the long run we come back to the question of what we mean by animal suffering.

In this essay I have come to the conclusion that the concept of suffering could usefully apply to an aversive, declarative state of the animal, but not to a state that is part of a routine behavioural procedure. The problem is how can we know whether, and when, an animal is in a state of suffering. This is a difficult problem, because it involves identifying a declarative state. All I can suggest, at this stage, is that evidence for declarative states are most likely to be found in animals in limbo, and in certain types of animal communication.

CHAPTER THREE

Economic altruism

In the evolutionary context, altruistic behaviour increases the individual fitness of the recipient and decreases the individual fitness of the donor. Natural selection may favour such behaviour under two main circumstances:

(a)   If the benefit in fitness to the recipient of an altruistic act exceeds the cost (decrement in fitness) to the donor by more than their coefficient of relationship.

(b)   If one individual benefits another at little cost and this situation is later reciprocated.

The latter form of altruism can occur between unrelated individuals, and is called reciprocal altruism (Trivers, 1971). Altruism towards kin can be regarded as selfishness on the part of the genes responsible (Dawkins, 1976), because copies of these genes are likely to be present in relatives. The evolution of this form of altruism poses no problems, provided individuals are able to recognize their kin, either individually, or by virtue of their circumstances or location. The evolution of reciprocal altruism does pose some problems, because individuals that cheat, by receiving but never giving, will be at an advantage. Various mechanisms for countering cheating have been suggested, including recognition of individuals and some form of policing (Trivers, 1985). For natural selection to favour reciprocal altruism, the individuals must have sufficient opportunities for reciprocation, they must be able to recognize each other individually and remember their obligations, and they must be motivated to reciprocate. These conditions are found in some primitive human societies, and it has been suggested that reciprocal altruism has played an important role in human evolution (Trivers, 1971, 1985).

In evolutionary terms, the benefits and costs of altruistic behaviour are increments and decrements in fitness. At the level of the individual animal, however, the costs are usually to be seen in terms of time and energy. In other words, the donor spends time and energy on altruistic behaviour, which could have been spent on some other behaviour of more direct benefit to the donor. In species with complex societies, altruism is difficult to quantify, because the relationship between fitness and the expenditure of time and energy is indirect. Thus an altruistic act may take the form of baby minding, loan of tools, or a political favour. In advanced human societies it may be possible to quantify these favours in terms of money, but this possibility is not evident in animal and human societies which have no money. On the other hand, it seems likely that money had its origins in exchange of favours, and some forms of economic altruism must have existed before money was invented. It seems likely that there is a continual thread linking the altruism of animals to that of modern human societies. The problem is to trace the thread.

Human and Animal Economics

Economics is usually regarded as the study of those activities which involve exchange transactions among people. Prominent amongst these activities is that of choosing how to allocate scarce resources among a variety of alternative uses. Economics is normally considered as a human activity, but many of its attributes also occur in other species. So we should perhaps reserve judgement about the definition of economics until we have carried out a comparison of human economic behaviour with that of other species.

Economists usually distinguish between macroeconomics and microeconomics. The former can be defined as the aggregate performance of all the parts of the economy, including the aggregates of income, employment and price levels. Microeconomics is the study of the particular incomes and prices that go to make up the aggregates. It is especially concerned with the choices made by individuals between work and leisure, and among various economic goods. The biological counterparts of macro- and microeconomics are ecology and evolutionary theory, on the one hand, and ethology and behavioural ecology on the other (McFarland and Houston, 1981).

A global economy should be organized in such a way that the maximum possible value is attained. This usually requires the allocation of scarce resources, such as food, time, and money, among competing ends. However, different opinions as to what is of maximum value will usually mean that different competing ends are considered to be of prime importance.

The concept of value itself is often a matter of political ideology. Thus to some (e.g. socialists) maximum value may be the equivalent of the greatest happiness for the greatest number; to others (e.g. free enterprise capitalists) it may be the greatest freedom of choice; to yet others (e.g. fascists) it may be the greatest good for the state. In some economies, therefore, the resources will be allocated amongst individual people, while in others the resources will be allocated among commercial companies, or departments of the state.

Economic principles apply not only to humans but to all animal species. All animals have to allocate scarce means among competing ends. The scarce means may be energy, nutrients or time. The competing ends may be growth and reproduction, as in a life-history study, or one activity versus another, as in a study of everyday decision-making.

At the global level, where economists think in terms of value biologists account for behaviour in terms of fitness. Just as some aspects of human economic structure or behaviour may be said to have greater social value than others, so some aspects of animal physiology or behaviour have greater fitness than others. Thus there is a parallel between the basic concepts of economics and biology, as illustrated in Figure 3.1.

At the microeconomic level, the level of individual decision-making, money is spent in such a way that utility is maximized. Utility is a notional measure of the psychological value of goods, leisure etc. It is notional because we do not know that it is involved in the choice process, only that people behave as if utility is maximized. Thus I may obtain a certain amount of utility from sport, from reading books, from collecting matchboxes etc. In spending my time and money on these things, I behave in a way that maximizes my psychological satisfaction, or utility.

Many biologists believe that animals behave in such a way as to maximize, or to minimize, some entity, sometimes called the cost. This is a notional measure of the contribution to fitness of the state of the animal and of its behaviour. Note that I say notional measure. The animal has no information as to the real cost. It only has a preprogrammed notion of the cost. I do not mean by this that the notion is cognitive, but merely that the information is somehow built into the animal. There is a fairly obvious parallel between cost and utility, which is illustrated in Figure 3.2. The function that is maximized by rational economic man is generally called the utility function. The equivalent function in animal behaviour is the objective function.

61

Economics	Ethology
(Meme)	Gene
Value	Fitness
Macroeconomics and sociology	Population genetics and sociobiology
(Smith, 1776) (Marx, 1867) (Keynes, 1936)	(Darwin, 1859) (Fisher, 1930) (Hamilton, 1963)
Utility	Cost

Utility function	Objective function
function that is maximized by individual person	function that is maximized by individual animal
(Policy function)	Cost function
social value will be maximized, in a given environment, if the person's utility function equals the policy function	fitness will be maximized in a given environment, if the animal's goal function equals the cost function

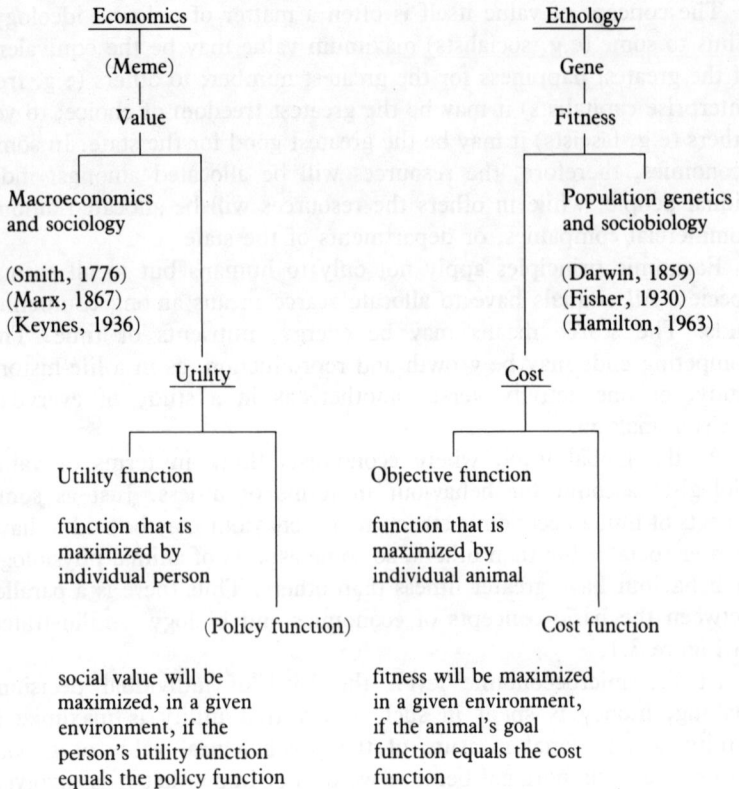

Fig. 3.1 Parallel concepts in economics and ethology. The names in parentheses are those of the most important contributors (with the dates of their major works) to microeconomics and population biology (after McFarland and Houston, 1981).

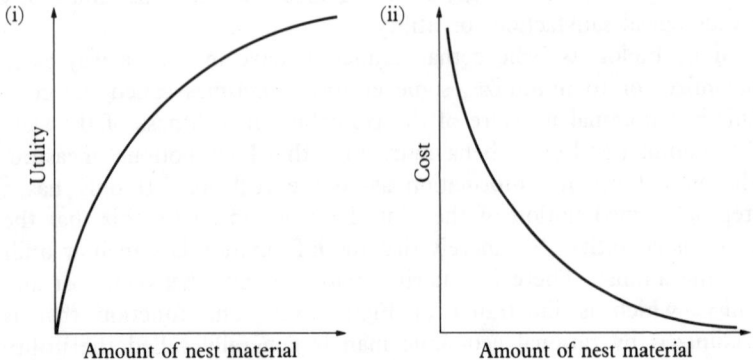

Fig. 3.2 (i) The decreasing marginal utility of nest material, and (ii) its equivalent cost (after McFarland and Houston, 1981).

We must distinguish between the function that the animal actually maximizes, the objective function, and the function that it should maximize if it were perfectly adapted to the environment. The latter function, called the cost function, is essentially a property of the environment rather than the individual. If an animal were perfectly adapted to its environment, then the objective and cost functions would be identical. In reality, animals are rarely perfectly adapted because of evolutionary lag, genetic variation between individuals, and competition between individuals (McFarland and Houston, 1981).

In economics it appears that the term utility function is used to denote both the function maximized by the individual and the function that the individual should maximize to attain maximum social value in a given environment. The latter may be more usefully called a policy function. In everyday life we are continually under pressure to change our utility functions so as to make them conform with some currently fashionable policy function. As children we may be pressed by our parents to give up socially undesirable habits. As adults we are pressed by governments to reduce the utility of smoking, or other activities considered to be contrary to current public health policy.

In likening an animal to an economic consumer, a distinction should be made between the common currency of decision-making, which is utility or cost, and the price or energy cost of behaviour (McFarland and Houston, 1981). Although the energy cost is sometimes a common factor, in the alternative, possible activities open to an animal, this is not always the case. It is a mistake, therefore, to regard energy as a common currency for decision-making, even though it may be a common factor in particular cases. Energy will be a factor common to all activities only if energy availability acts as a constraint in the circumstances being considered. An everyday economic example may help to make this distinction clear.

Suppose a person enters a supermarket to purchase a variety of goods. Associated with each possible item of purchase will be a utility function, the utility of particular goods depending on the shopper's preferences and experience. The concept of utility makes it possible to envisage how very different items are compared. A person may obtain the same utility from a packet of soap as from a bottle of soy sauce. Utility is thus the currency by which different items are evaluated. However, another currency common to all items is money. If the shopper has a limited amount of money to spend, then money acts as a constraint on the amount of goods that can be purchased. However, money is not always the relevant constraint. A shopper in a hurry may carry more money than can be spent in the time available. It may turn out that the shopper's choice when pressed for time is different from that of the shopper with limited cash.

There are many possible constraints that may impinge upon an animal's choice of behaviour. These may vary with circumstances, but it is only the constraint that bites at the time that is important. For this reason, time and energy should not be confused with utility or cost. In drawing the analogy between an animal and an economic consumer, cost is equivalent to utility, energy is equivalent to money, and time plays the same role in both systems.

The animal can earn energy by foraging, just as a person can earn money by working. The energy, or money may be spent upon various other activities. Over and above the basic continuous level of metabolic expenditure, or subsistence expenditure in the case of people, the animal can save energy by depositing fat and hoarding food, and the person can save money by banking it, or putting it in a sock. Alternatively, the animal can spend energy upon various activities, including foraging, while the person can spend money on goods or activities, including working.

As we saw in Chapter 2, the amount purchased by a shopper depends partly upon the price. For example, when the price of coffee is increased in the supermarket, people continue to buy about the same amount as before, perhaps a little less. As the price of fruit is increased, however, demand for fruit falls off. When the price of fresh fish is increased demand falls markedly. Presumably, people are willing to pay more to maintain their coffee-drinking habits. Demand for coffee is said to be inelastic. If the price of fish increases, however, people tend to buy less and to switch to substitute foods such as meat or canned fish. Demand for fish is said to be elastic.

Exactly analogous phenomena occur in animal behaviour. If an animal expends a particular amount of energy on a particular activity, then it usually does less of that activity if the energy requirement is increased. Numerous studies have shown that demand functions in animals follow the same general pattern as those of humans. Examples cover a wide variety of animals, ranging from humans to protozoa, performing many different kinds of responses as behavioural payment for a wide variety of commodities, including food, water, sucrose, saccharin, heat, cigarettes, alcohol and various drugs (Allison, 1983). Figure 3.3 shows some examples.

When a commodity (or activity) increases in price, the consumer (or animal) has an effectively reduced income. The consumer may buy less of the commodity for the same money (a case of elastic demand), or pay more for the same amount of the commodity (a case of

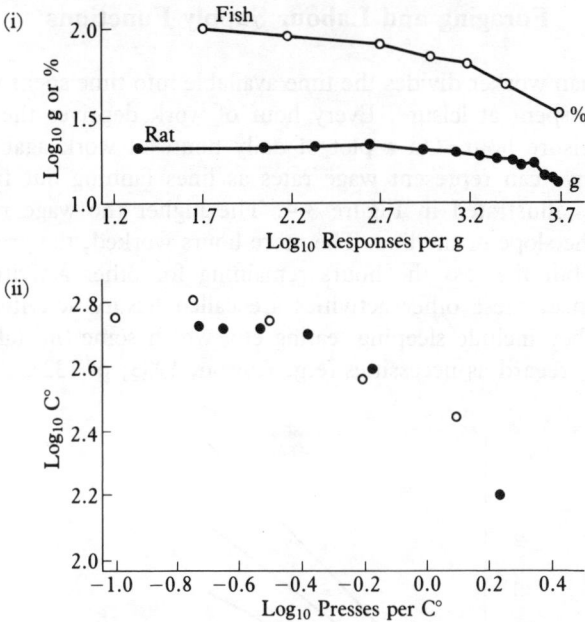

Fig. 3.3 Examples of demand functions in animals. (i) Total food consumed by rat or fish as a function of the behavioural price of food. (ii) Heat consumption by rats as a function of the behavioural price of heat (after Allison, 1983).

inelastic demand). In the latter case, the consumer will often have less money to spend on other goods. Thus an increase in the price of a commodity can be thought of as a decrease in the consumer's income under conditions of constant prices. This is known as the income effect. Another result of one commodity becoming more expensive is that other commodities become relatively cheaper. If demand for the commodity whose price has risen is elastic, then the consumer buys less of that commodity and has more to spend on other commodities. This is known as the substitution effect. The demand for a commodity whose price has risen is usually elastic when there are good substitutes available.

The substitution effect has been shown to occur in animals. For example, Lea and Roper (1977) found that demand for food was inelastic in rats required to work (press a lever) to obtain food rewards. As the work required (lever presses per reward) increased, the rats continued to work for the same amount of food. However, these investigators also found that elasticity of demand for food pellets increased when sucrose was available as a substitute.

Foraging and Labour Supply Functions

The human worker divides the time available into time spent working and time spent at leisure. Every hour of work deprives the worker of one leisure hour. On a plot of daily hours of work against total income, we can represent wage rates as lines fanning out from the origin, as illustrated in Figure 3.4. The higher the wage rate, the greater the slope of the line. The more hours worked, the greater the income, but the less the hours remaining for other activities. For convenience, these other activities are called leisure activities, even though they include sleeping, eating etc, which some (mistakenly in my view) regard as necessities (e.g. Allison, 1983, p. 132).

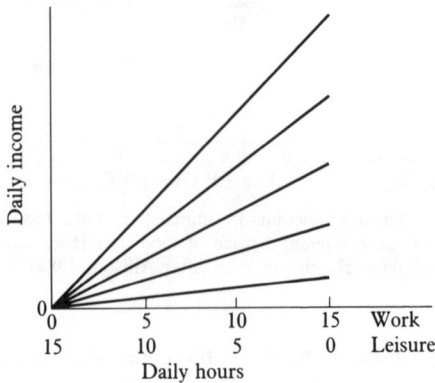

Fig. 3.4 Daily income as a function of wage rate, daily work hours and daily leisure hours (after Allison, 1983).

Assuming that a worker is free to choose the amount of work done each day, we can expect that there will be a trade-off between work and leisure, resulting in a set of indifference curves tangent to the wage rate lines, as illustrated in Figure 3.5. All points on a particular indifference curve have equal utility for the worker. The shape of these curves suggests that income and leisure are imperfect substitutes (see above). Thus the worker with a high income will sacrifice a large amount of money to obtain increased leisure time, while the worker on a low income will sacrifice little.

The optimum number of working hours, at a particular wage rate, is given by the point on the line which attains the greatest utility; that is the point at which the highest possible indifference curve is reached, as illustrated in Figure 3.6. By joining the optimum point

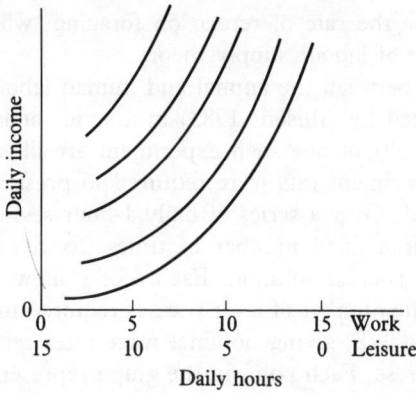

Fig. 3.5 Iso-utility (indifference) curves for money and leisure as imperfect substitutes (after Allison, 1983).

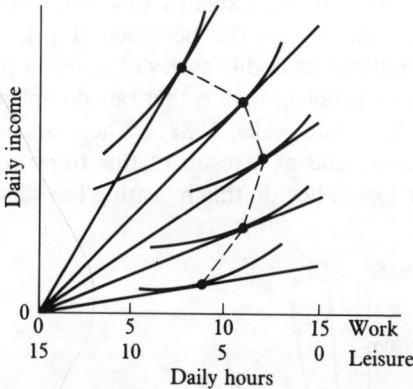

Fig. 3.6 A backward-bending labour supply curve (dotted line) (after Allison, 1983).

on each wage rate line (by a dotted line) we attain the labour supply curve of economic theory.

Animals spend time and energy to obtain food, just as people who earn money by working must expend time and some money while working. The animal equivalent of money is energy. Thus animals earn energy by foraging, and spend it on other activities. Different species employ different methods, some searching for food, some lying in wait for prey, some grazing etc. Thus some species have a high rate of energy expenditure while foraging, but spend relatively little time foraging, while other species have a low rate of energy expenditure but spend a lot of time foraging. The rate at which energy is obtained depends upon the availability and accessibility of the food.

These determine the rate of return on foraging, which is analogous to the wage rate of labour supply theory.

The analogy between the animal and human labour supply curves has been explored by Allison (1983) in a series of laboratory experiments. The results of one such experiment are illustrated in Figure 3.7. In this experiment rats were required to press a lever to obtain a sucrose reward. Over a series of daily 1-hour sessions each rat had to press the bar a fixed number of times (fixed ratio schedule) to obtain access to sucrose solution. Each access allowed 10 licks at the sucrose tube. The number of lever presses required to obtain a reward varied from 1 to 128, giving nominal wage rates ranging from 10 to 0.08 licks per press. Each point in the graph represents the total licks (the analog of income) obtained for total lever presses (the analogue of labour) at one of the eight wage rates. Notice that the analogue of the labour supply curve bends back in a manner similar to that illustrated in Figure 3.6. Allison explains this phenomenon in terms of his 'conservation' theory. As the behavioural price of water grows high, the rat substitutes its bodily stores of water in place of the costly external water. One problem here is that rats do not really have bodily stores of water (Rolls and Rolls, 1982). When rats are short of water they become thirsty, and as a result of this thirst they cut down on the rate of water loss. They do this by eating less (and thus obligating

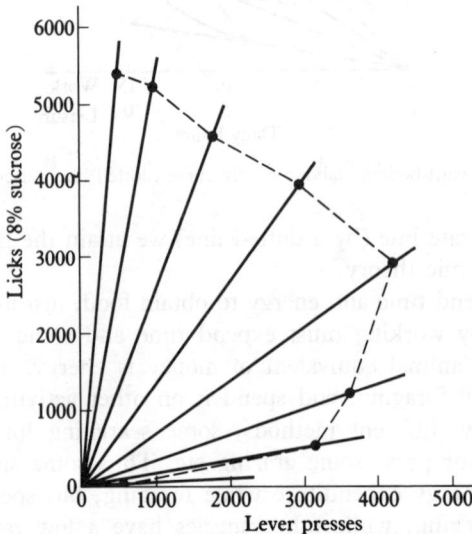

Fig. 3.7 Total sucrose licks earned by rats as a function of total lever presses at eight different wage rates (licks/press) (after Allison, 1983).

a lower water loss in the excretion of waste products), and by reab-sorbing water in the kidney and small intestine. These processes incur physiological costs, because they result in lower energy intake and accumulation of waste products. In drinking less water when the behavioural cost of water is high, the rat is involved in a set of complex physiological trade-offs. These can be formulated in terms of an optimality theory (McFarland and Houston, 1981), but all I can say here, without digressing far from my overall theme, is that I doubt that simplistic behavioural 'conservation' models are capable of hand-ling issues of this complexity. Instead, I propose a much more straightforward explanation of the back-bending labour supply curve.

If the animal has some other source of energy income (or water income), apart from foraging, then the wage-rate line does not pass through the origin, but intercepts the ordinate, as illustrated in Figure 3.8. The extra income may come from fat reserves, hoarded food, stolen food, food given by other animals etc. The effect of such income will generally be to reduce the time spent foraging. Indeed, we can use such knowledge to sketch out the labour supply situation for an hypothetical animal.

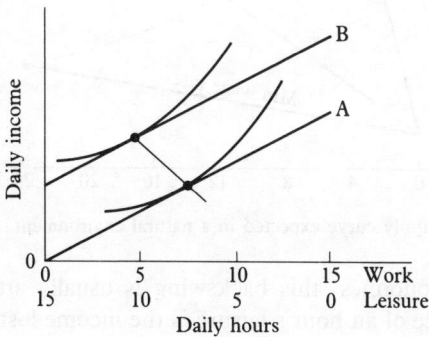

Fig. 3.8 Daily income as a function of work, leisure and non-labour income (after Allison, 1983).

There is quite a lot that we can say *a priori* about the labour supply situation in animals. Firstly, there must be a sufficient energy income to sustain the basal metabolic rate. So we can assume that an animal would be prepared to work nearly 24 hours a day at the lowest sustainable wage rate, otherwise it will die. There may be some animals willing to sacrifice their lives for some activity other than foraging, but these must operate above the minimal wage rate in order to have energy available for these other activities. Secondly, we can expect that there will always be a maximum income, earned or otherwise, beyond which the animal is simply not interested. We cannot expect to find animal millionaires who accumulate energy simply for the sake of it. The maximum income is that beyond which the animal derives no utility from further income.

Thirdly, we can assume that our animal will forage less when given free food, as has been documented in some cases (see below). This means that the labour supply curve must move downwards from left to right, as illustrated in Figure 3.8. Taken together, these three points suggest a labour supply curve such as that illustrated in Figure 3.9. Notice that we do not end up with the backswing illustrated in Figures 3.6 and 3.7.

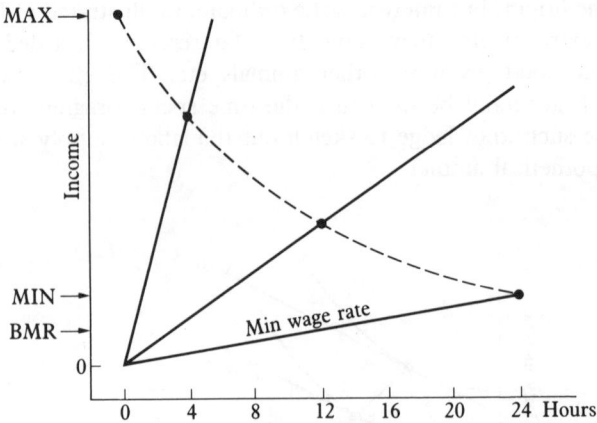

Fig. 3.9 Labour supply curve expected in a natural environment.

In human economics, this backswing is usually attributed to the fact that the price of an hour's leisure is the income lost from the hour not worked. As the wage rate is reduced work becomes less worthwhile and leisure becomes cheaper. We must remember, however, that in human economics this phenomenon occurs in highly developed, rather than peasant communities. Those people who work less as wages drop usually have a hidden income in the form of social security. We must consider, also, the circumstances under which Allison's experiments were probably conducted. The sucrose obtained during the experiments was almost certainly not the rat's only source of daily food. They were probably already on a high free income. Moreover, for the 23 hours per day that they were not involved in the experiment they were probably couped up in solitary confinement,

70

as is usual in most psychology laboratories. Their leisure was cheap in energy terms, because there was nothing to do. In the natural environment, however, some of the activities that animals are involved in when not foraging require a lot of energy. There are some special circumstances in which animals cut down on feeding, such as incubation in junglefowl (Sherry, Mrosovsky and Hogan 1980) and rut in red deer stags (Clutton-Brock, Guinness and Albon, 1982), but these animals have an income from stored food. I suggest, therefore, that, while foraging may be cut back when wage rates are low, under conditions of high income, this does not occur when income is low. The true situation, as I see it, is illustrated in Figure 3.10.

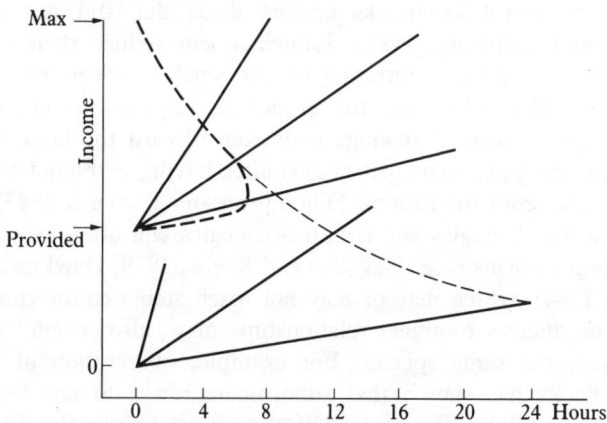

Fig. 3.10 Labour supply curve obtained when some non-labour income is provided in a natural environment (assuming iso-utility functions are the same as those implied in Fig. 3.9).

Energy Income, Expenditure and Altruism

Food is the prime source of energy in animals. Food may be gained in various ways, all of which may be listed under the heading of foraging, as shown in Figure 3.11. Natural selection favours efficient foragers, and many animals are extremely adept at searching for, and harvesting food. However, the food-finding abilities of one species can become part of the foraging strategy of another species. In other words, the foraging investment of one species (the producer) is parasitized by another species (the scrounger). For example, lapwings foraging on earthworms are pirated by black-headed gulls. Often the gulls are involved in expensive aerial chases. They can avoid these if

71

```
                      Forage
           ┌────────────┴────────────┐
        Produce                     Usurp
     ┌─────┴─────┐            ┌───────┴────────┐
   Farm        Search     Parasitize       Manipulate
```

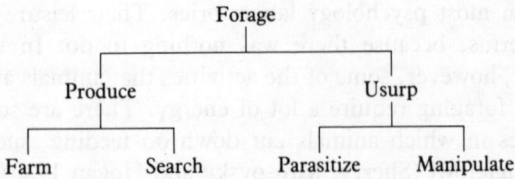

Fig. 3.11 Modes of food gain in animals.

they attack the lapwings just as they obtain their prey. When taking large worms, the lapwings adopt a tell-tale crouching posture, which the gulls can use as a cue as to the best time to attack. In one study it was observed that attacks against crouching lapwings were four times as successful as attacks against those that did not crouch (Barnard and Stephens, 1981). Lapwings can reduce their vulnerability to attack by concentrating on the smaller worms nearer the surface, and thus obviating the crouching required to obtain the larger, deeper worms. Lapwings may even discard the large worms (which are the gulls' principal target) after having extracted them at great expense from the ground (Thompson and Barnard, 1983). The kleptoparasitic strategies and counter-strategies will often give rise to an evolutionary arms race (Dawkins and Krebs, 1979; Dawkins, 1982; Barnard, 1984), which may or may not reach stable equilibrium.

The producer-scrounger relationship may also occur among members of the same species. For example, observation of Harris sparrow flocks has shown that subordinate birds do not feed less efficiently, as might be expected if they were frequently displaced from the good feeding areas. The dominant sparrows sometimes allow a few subordinates to feed with them and indicate the whereabouts of food. The subordinates may be defended against use by other dominants, and thus gain some advantage over other subordinates. At the same time some of the food found by the subordinates is stolen by the dominant bird (Rohwer and Ewald, 1981).

Some animals adopt an alternative foraging strategy when this is less costly (in terms of fitness) than the normal strategy. The best strategy may depend upon the strategy adopted by others, and in some cases, a mixed strategy (ESS) will result. Many of these considerations have been the subject of formal modelling (e.g. Sibly, 1984), and a number of animal species have been studied from this point of view (see the many examples in Barnard, 1984).

Once gained, food is disposed of in one of the five ways illustrated in Figure 3.12. Food that is eaten and digested makes a certain amount of energy available. Some of this is wasted in the process of

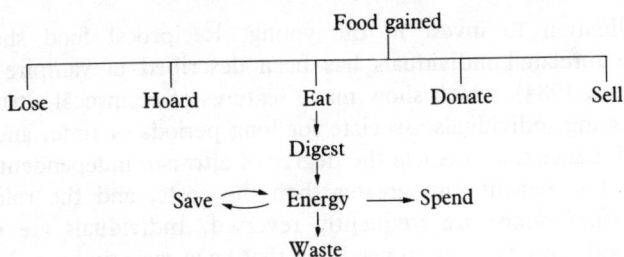

Fig. 3.12 Modes of food disposal in animals.

excretion, some is stored in the liver or as fat deposits, and the rest is spent on various physiological and behavioural processes. Not all food that is obtained by foraging is immediately eaten. Some is hoarded, some donated to others, some sold and some wasted. Let us discuss these in order.

First, hoarding is essentially a behavioural method of storing energy. Many species hoard food that is surplus to their immediate requirements, though hoarding practices vary widely among species. For example, it appears that hamsters respond to food scarcity, not by foraging more as do rats and gerbils, but by increasing the rate of consumption of hoarded food and decreasing the rate at which they add to the hoard. They respond to high food availability by greatly increasing their rate of hoarding (Lea and Tarpy, 1986). Thus they treat their food hoard in the same way that other species treat their fat deposits.

Secondly, donation of food by one individual to another is common amongst animals. The most obvious example is feeding the young. This is a form of parental investment (Trivers, 1972, 1985) and can readily be explained in terms of kin selection. More difficult to account for is food donated to the young, not by parents, but by helpers. Amongst birds and mammals it is usually relatives of the parents who help to raise the offspring (for reviews see Trivers, 1985; Emlen, 1984). The donation of food to non-relatives occurs as courtship feeding in many species of insects, spiders and birds. The male presents food to the female. In most species the quantity of food transferred from male to female is small in relation to the female's normal daily intake. In the herring gull the female begs for food as a juvenile would, and the male regurgitates food, which the female then eats. Niebuhr (1981) found that courtship feeding was a reliable predictor of chick feeding by the male herring gull. This finding reinforces the common view (e.g. Trivers, 1985; Nisbet, 1977) that courtship feeding is demanded by the female as a test of the male's ability and inclination to invest in the young. Reciprocal food sharing amongst unrelated individuals has been described in vampire bats (Wilkinson, 1984), which show many features of reciprocal altruism. Reciprocating individuals associate for long periods of time, and the degree of association predicts the degree of altruism independently of kinship. The benefits are greater than the costs, and the roles of recipient and donor are frequently reversed. Individuals are more likely to aid those that are in need and that have recently been donors themselves.

Thirdly, food gained may be sold. That is the food may be exchanged for money, or some equivalent of money. Amongst humans this practice is commonplace, and the exchange can take place immediately, or payment can be delayed. Amongst animals, the equivalent of money is energy, so the question is: can food be given by one individual to another in exchange for energy? Since food is the only way of transferring energy between individual animals (except perhaps for the donation of bodyheat), the answer to this question must be in the negative, at least if we are thinking of immediate exchange. What about delayed payment, however? Can food be donated by one individual in exchange for future payment in terms of energy (food). Obviously, this is little different from the type of reciprocal altruism just described for vampire bats. So it seems that selling food may have something in common with reciprocal altruism.

Fourthly, food gained may be lost; as mentioned above, some birds, such as sparrows and lapwings, are pirated by other birds. This is less likely to happen when the forager eats the food soon after it is obtained, but it is quite common (Barnard, 1984) when the forager is transporting food to donate it to others (usually its young). Food may also be lost through wastage. A predator may not be able to eat all that it has killed, or recover all that is has hoarded.

It is difficult to pick out those factors that distinguish human foraging from that of other species. Hunting, piracy, gathering and 'farming' occur in a number of species. Animals often consume their food as soon as it is obtained, but they may also hoard it or donate it to others, as we have seen. Humans may sometimes eat food as soon as it is obtained, but they do not usually live a hand-to-mouth existence. Perhaps it is exchange that distinguishes humans from other species.

Among humans we find subsistence living of various different types. Historically the most primitive is usually regarded as the hunter-gatherer existence, based on wild plants and animals. Primitive agriculture leads to domestication of plants and animals, and to more complex economic organization. The products of human labour are distributed by means of exchange, except when working individuals consume their own produce. Exchange is the practice of giving and receiving valuable objects and services. The term is traditionally confined to human behaviour, although, as we have seen, donation of food and other materials is common in the animal kingdom.

A summary of the different types of exchange, as classified by Harris (1985), is given in Figure 3.13. Reciprocal exchange is characterized by a giving of labour products and services that is not contingent on any definite receipt of goods and services in return. Rather, the giving is based on a general understanding that there will be some eventual reciprocity. As we shall see, this form of exchange is characteristic of some hunter-gatherer societies. Some form of reciprocal exchange occurs in all human societies, especially among relatives and close friends. Thus informal transactions of a minor sort are commonplace within the family, on an everyday basis. A more ritualized reciprocal exchange of presents also occurs on more formal occasions.

Reciprocal exchange is open to abuse by cheats, or free-loaders, who receive favours from others, but do not reciprocate. In some cases, such as the giving of food, clothing etc., to children, no reciprocation is expected, but a roughly symmetrical reciprocity is expected in exchanges amongst adults. In societies practising reciprocal

Means of exchange	Definition	Control	Social	Ecology
Reciprocal exchange	Whoever produces shares	Cheats disliked	No thanks given or expected	Production subject to diminishing returns
Egalitarian redistribution	Whoever produces gives	Accounts kept	Producers boast and receive gratitude	Productivity encouraged
Stratified redistribution	Whoever produces must contribute	Policed	Political control of producers	Productivity controlled
Barter market	Producers bargain	Accounts kept	Some trust required	Few commodities
Price market	Producers sell	All-purpose money	Trade with strangers	Many commodities in mixed economy

Fig. 3.13 Summary of modes of exchange in humans.

exchange, free-loaders are subject to subtle forms of disapproval, but demands are not made upon them in particular cases. It is usually part of the ethos of reciprocal exchange to deny that any balance is being calculated. Those that provide do not boast, or expect to receive thanks, as Harris (1985, pp. 235–6) explains:

> Boastfulness and acknowledgement of generosity is incompatible with the basic etiquette of reciprocal exchanges. Among the Semai of Central Malaya. no one even says 'thank you' for the meat received. . . Having struggled all day to lug the carcass of a pig home through the jungle heat, the hunter allows his prize to be cut up into exactly equal portions, which are then given away to the entire group . . . to express gratitude for the portion received indicates that you are the kind of person who calculates how much you are giving and taking. . . To call attention to one's generosity is to indicate that others are in debt to you and that you expect them to repay you. It is repugnant to egalitarian peoples even to suggest that they have been treated generously.

In societies more complex than family, or hunter-gather, groups some form of coercive exchange is the norm. Redistributive exchange involves a small amount of coercion, and has much in common with reciprocal exchange. The produce of several individuals is brought to a central place, sorted and counted, and shared among producers and non-producers alike. On a small scale, this is little different from reciprocal exchange, but some extra organization and co-ordination is required if large quantities of material are to be collected in one place, and distributed at the same time. The organizational role is sometimes taken by a prominant provider who has an interest in increasing production, and in receiving recognition for his efforts. This is called egalitarian redistribution, and it differs from reciprocal exchange primarily in encouraging productivity.

In other cases the redistribution is done by an individual who takes no part in the production, and usually retains the largest share of the produce. To fulfil such a role the redistributor must be able to exercize some form of coercion, or political power. The society is no longer egalitarian, but stratified. A certain amount of control over productivity is characteristic of stratified redistribution.

Harris (1985) notes that reciprocal exchange, characterized by modesty on the part of the producers, tends to occur in ecological situations in which increased productivity would tend to produce diminishing returns. Hunter-gatherer societies cannot afford to increase their predation beyond the point of ecological balance. Agricultural peoples, however, can often benefit from increasing production without endangering their food supplies. They can afford to admire and encourage those who are big providers.

Exchange amongst unrelated (non-kin) groups usually takes place at some kind of market place. In the absence of money, goods and services are bartered. There is bargaining between the opposing parties, each trying to maximize their profitability. If the barter exchanges are not direct, then some sort of accounts must be kept. For this reason, and because of the danger of possible use of force by one, or both, parties, there must either be a degree of mutual trust, or some form of policing. The barter market can work satisfactorily if relatively few commodities are involved in the exchange.

A society which has some form of all-purpose money can operate a price market. The producers sell their goods in exchange for money, and use money to purchase other goods or services. The main advantage of the price market is that it can handle many commodities in a mixed economy. Another advantage is that it facilitates trade with strangers, with whom no elaborate trust arrangements exist. Traditionally, money has been regarded as a portable, recognizable material. It has a certain legality which makes it divisible and convertible, and confers wide generality of use. In recent times, however, we have started to move away from money as a physical object of exchange. Bank accounts and credit cards enable us to promise each other that we will pay our debts at some time in the future. In this respect we seem to be returning to some of the features of reciprocal exchange.

To investigate the relationship between energy and money we should look at human societies that have no money. Unfortunately, few such societies remain unaffected by modern cultural influences. However, some such have been studied by anthropologists at a time when such influences were less than they are today. Richard Lee's studies of the !Kung San are a good example. These took place between 1963 and 1973 in the Dobe area of the Kalahari desert, Botswana, where there were some relatively independent !Kung groups. Most of the information that I have used comes from Lee (1979).

The !Kung San of the Kalahari desert provide a good example of the importance of reciprocal altruism in hunter-gatherer societies. The men spend many hours hunting for game. The food supply is very variable and periods of plenty may be followed by hard times. The men usually hunt in pairs or small bands, and if one band is successful, the meat is shared among all members of the group. The successful provide food for the unsuccessful on the understanding that the situation may be reversed. The successful never boast about their achievement, but are always very modest and self-effacing. As we have seen, this is typical of hunter-gatherers and contrasts with the boastful behaviour of people who practise redistributive exchange. Harris (1985) maintains that the typical reciprocal exchange occurs in environments where hunting follows the law of diminishing returns, over-exploitation of the game is not to be encouraged, so the different hunting groups do not compete, and do not advertise their prowess.

The women bring in about 60 per cent of the protein and carbohydrate by gathering vegetables and fruit. While foraging for plant foods they may be severely encumbered by their children. Mongongo nuts, their most important plant food source, are in plentiful supply in the dry season but are usually situated some six miles from suitable campsites. The women make excursions to gather these nuts every few days, taking their children with them. Lee (1972) shows that the weight carried by mothers on these excursions increases with the frequency of having babies, not only because there are more mouths to feed but also because the small children have to be carried. Blurton-Jones and Sibly (1978) calculate that the average birth spacing of four years is the optimum under the prevailing conditions. When studied by Lee, the hunter-gatherer population was expanding. The population density remained static, but there was considerable out-migration.

It is interesting to compare the plants eaten by the San with those used by the Gwembe Tonga, studied by Scudder (1971) in Zambia. The Tonga are an agricultural people with a population density 100 times greater than the San. The Tonga use about 131 species of wild plants compared with 105 species for the !Kung San. The Tonga use the wild plants as emergency food during times of crop failure. Moreover, they consume portions of 21 species that are available to the !Kung but not regarded as food. The pressure put on the wild flora by the Tonga in times of need is very much greater than that exerted by the !Kung, even in times of drought. It seems paradoxical that an agricultural people should use their plant and animal resources more intensively than a hunting and gathering people. Lee (1979) concludes that the foraging way of life is a fairly secure one.

The !Kung San have plenty of available food provided that they keep moving. The longer they stay in one place, the farther they have to walk to obtain food. Eventually the camp location is no longer viable, and the whole group moves to a new location.

An Economic Model

The foraging members of the group work to obtain calories which are shared among all members of the group, plus any visitors. The consumption of each person, therefore can be assumed to be the same, in relation to their requirements. The average basic metabolic rate (BMR) BMR for women (1100 kcal) is less than that of men (1400 kcal). Similarly, the working requirements for women (1750 kcal) are less than for men (2250 kcal). We can use these requirements to set up a preliminary model of the labour supply situation.

The San do not have any social security, so we must assume that their labour supply situation is something like that illustrated in Figure 3.14. This is a version of Figure 3.9, scaled according to the San situation as outlined by Lee (1979). Here we have a calorie scale of income, on which the lower limits of individual survival are given by the figures for basal metabolic rate. If we average the values for men and women (for the time being), then the minimum wage rate that permits survival is OZ (assuming that foraging is restricted to the 12 hours of daylight).

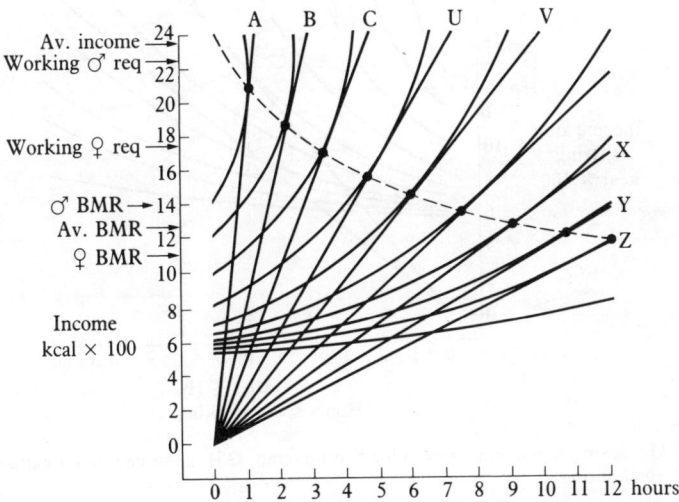

Fig. 3.14 Constructing a labour supply curve for the San. The labour supply curve (dotted) cannot go below the average BMR; the wage rates must fan out from the origin, so the iso-utility curves must be something like those shown.

79

The San are nomadic, and do not store food, so the daily income is all consumed. The maximum daily intake must therefore be only a little more than the average daily requirement. On this basis the maximum wage rate is given by the line OA. If we assume a reasonably uniform pattern of iso-utility curves, then the labour supply curve for foraging San is going to be something like that illustrated in Figure 3.14. Note that this is similar to that illustrated in Figure 3.9, and is typical (I argue) of a situation in which there is no equivalent of social security. It might be argued that the San provide their own social security, since the old, sick, and idle receive their share of the food income. I will take up this point at a later stage.

We know (from Lee, 1979) that 60 per cent of the total calorie income comes from gathering wild plants. The average daily income is 2355 kcal, so the average daily income from gathering is some 1400 kcal, shown as wage rate G in Figure 3.15. The model indicates that it should take 6 hours per day to produce this income.

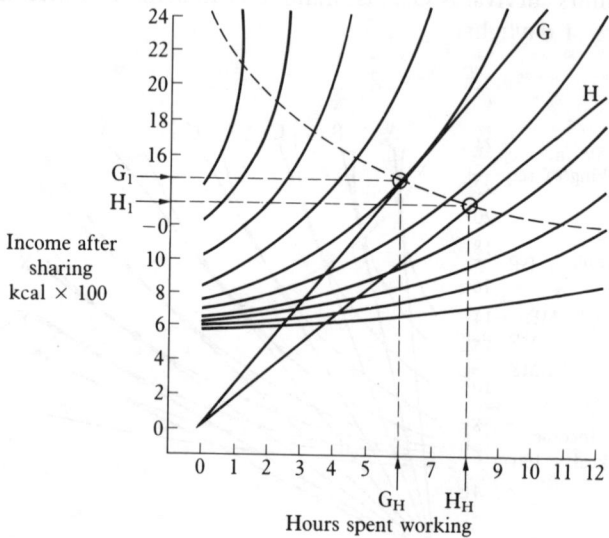

Fig. 3.15 A simple economic model for San foraging. G is the wage rate for gathering, H that for hunting.

A man-day of hunting brings in 7230 kcal on average, compared with 12,000 kcal for a person-day of gathering. So the hunting wage rate is only 60 per cent of the gathering wage rate. On this basis, the model predicts an 8-hour day for hunting, with an income of about

1250 kcal after sharing, as indicated by wage-rate H in Figure 3.15. As it happens, the women usually work a 6-hour day and the men an 8-hour day, when foraging. The hunting and gathering parties usually leave camp at the same time of the morning, but the women come home early to do camp chores (Lee, 1979). Thus our preliminary model appears to be on the right lines.

In gross terms, the productivity per hour spent foraging is very much greater than is shown here. The women procedure 2010 kcal/hour, and the men 1029 kcal/hour. However, this does not take account of the handling time. Most of the food obtained by hunting and gathering is transported to the camp, prepared, and shared out. Even if these people lived a hand-to-mouth existence, they would still have to prepare the food (mongongo nuts require a lot of work to crack) and maintain their tools in good order.

In terms of hours per week, the work load of men and women can be summarized by the chart illustrated in Figure 3.16. This shows that in 21.6 hours of foraging, a man will produce 22,221 kcal, but he works a 44.5-hour week in total, so he earns just under 500 kcal/hour. In 12.6 hours of foraging a woman earns 25,333 kcal, giving a 631 kcal/hour wage rate for a 40.1-hour week.

San work hours per week

	Foraging	Repairs	Homework	Total
Men	21.6	7.5	15.4	44.5
Women	12.6	5.1	22.4	40.1

Fig. 3.16 Summary of San hours of work per week (Data from Lee, 1979).

Whether foraging or doing camp chores, the individual is working for payment that is delayed. The payment comes in the form of shared-out food. If the payment had come in the form of money, it would have to be spent on that same shared-out food, because there is nothing else to buy. We can use these figures to calculate the notional wage earned by hunters and gatherers.

As we have seen the BMR for women is 1100 kcal, while the requirement for a working woman is 1750 kcal. Therefore, as a multiple of the BMR, the requirement is 1.59. Similarly, the BMR for men is 1400 kcal, while the working requirement is 2250 kcal, which is 1.60 as a multiple of BMR. In other words, the calorie requirement for working men and women is the same, when taken as the MBMR (multiple of the BMR). MBMR is a convenient unit

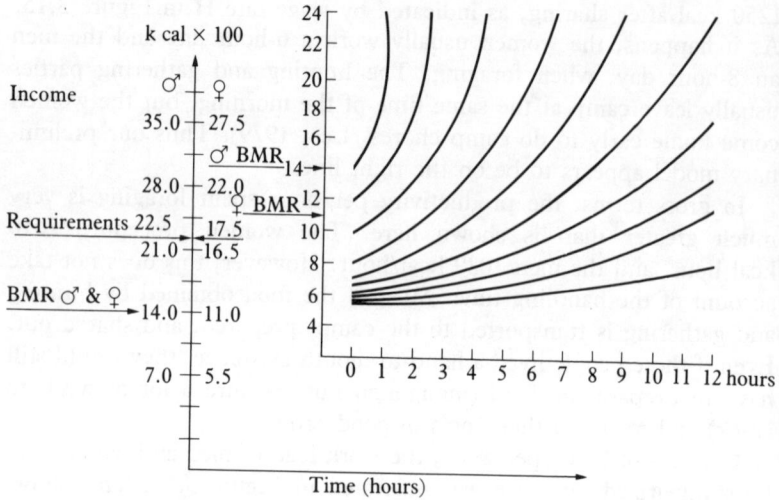

Fig. 3.17 Recalibration of San calorie income in terms of multiples of the basal metabolic rate (MBMR).

of comparison of individuals of different body size, and we can rescale our calculations in these terms, as illustrated in Figure 3.17.

In terms of MBMR the wage rates for men and women are illustrated in Figure 3.18. The men work a 44.5-hour week overall, which is equivalent to a 6.35-hour day. They earn 500 kcal/hour, or 3180 kcal/day, or 2.27 MBMR. A man spends 2250 kcal/day on himself, and so he has a net income of 930 kcal/day, or 0.66 MBMR. The women work a 40.1-hour/week, equivalent to a 5.73 hour/day. They earn 631 kcal/hour, or 3615.5 kcal/day, or 3.28 MBMR. A woman spends 1750 kcal/day on herself, so she has a net income of 1865 kcal/day, or 1.7 MBMR. In other words, a gathering woman is more than twice as productive (in MBMR terms) as a hunting man. Although we should remember that the food brought in by the men may be particularly valuable in nutritional terms.

Taking the working requirement for both sexes as 1.6 MBMR, we can set the minimum wage rate (Figure 3.18). At first sight, these data suggest that the labour supply curves of women differ from those of men. If this is so it means that the utility functions of men and women must be different. However, on the basis of Figure 3.15 it is possible to show that the utility functions of men and women are the same, when calculated in terms of MBMR. These functions give the iso-utility curves shown in Figure 3.19. The labour supply curve L must

82

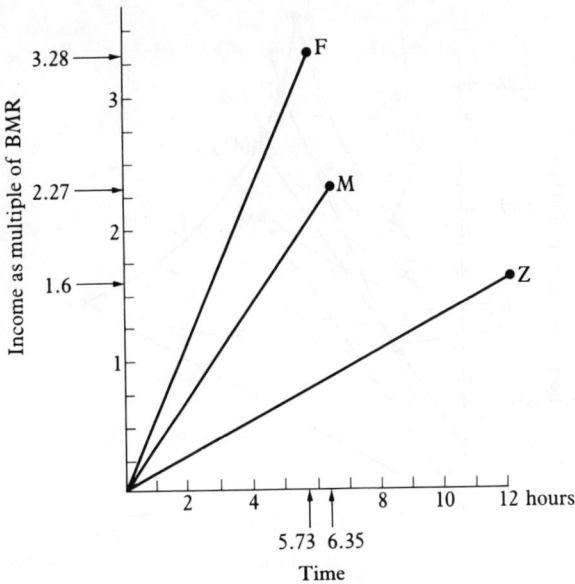

Fig. 3.18 San wage rates in terms of MBMR. F = female, M = male, Z = minimum wage rate.

pass through point F, the female income point (from Figure 3.18) and Z, the minimum income point. The male income point M (from Figure 3.18) is below this curve. We must remember, here, that we have not yet taken into account the fact that the income is shared. Let us assume, for the time being, that the income is shared among producers, before being distributed more widely. The female net income is 1865 kcal/day (1.7 MBMR, see above) and the male net income is 930 kcal/day (0.66 MBMR). Therefore, if each female gives up 635 kcal/day, her resulting net income is 1230 kcal/day = 1.118 MBMR. If each male receives 635 kcal/day his resulting net income is 1565 kcal/day = 1.117 MBMR, equivalent to the female resulting net income. One way of representing this partial sharing is to consider that the male receives a free income of 635 kcal/day or 0.45 MBMR. Accordingly, in Figure 3.19 we raise the male wage rate by this amount, and we find the male income point M^1 now rests on the labour supply curve. In fact this small amount of extra income makes little difference to the overall labour supply curve, and we can thus expect that working men and women will respond similarly to changes in wage rates.

83

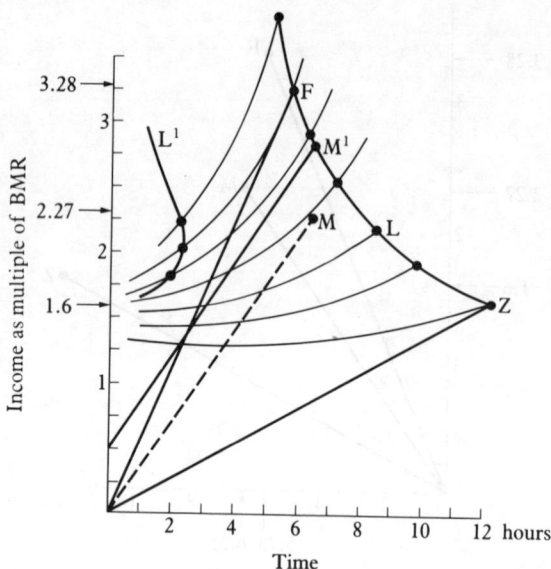

Fig. 3.19 Revised model of San hunting and gathering. L = basic labour supply curve. L' = labour supply curve of a man with a free income of 1.6 MBMR. M = male wage rate from Fig. 3.18. M' = male wage rate after donation of 0.5 MBMR from females.

Lee (1979) reports that some males never go hunting, and do virtually no work. Such men receive food from the ultimate sharing-out process, just like any other member of the community. If we assume that they receive the minimum income for a (working) man, then we obtain the labour supply curve L¹ in Figure 3.19. From this we can see that little work is done by those whose free income is sufficiently high. Here we have an apparent paradox. If an individual can survive comfortably without working, why do not more of them do so? Why does the group continue to support those that do not work? There are many possible answers to these questions.

Women may have more of an interest in sharing than do the men. This might be the case if there was some uncertainty about paternity. Another possibility is that the nutritional benefits of hunting have been undervalued in my calculations. In addition to the calories provided by the meat, there may be some essential nutrients not provided by plant material. A third possibility is that reciprocal altruism, as calculated purely in terms of calories, is assymetrical, possibly because one party reciprocates in some other way.

84

In terms of time, the men appear (Figure 3.16) to contribute slightly more than the women, but we must remember that these figures are somewhat rough and ready. The main difference between the sexes in one of productivity. The women are nearly twice as productive as the men, but the men could remedy this inbalance by spending more time gathering and less time hunting. If we assume that the women are incapable of hunting successfully, then what could the women do if the men gave up hunting and took up gathering? If both parties spent all their time gathering, there would be a considerable over-production of calories. As we have seen, the system of reciprocal exchange practised by the San is designed to avoid over-exploitation of the environment, and excessive productivity is discouraged.

It is not my purpose to offer an explanation for the apparent assymetrical reciprocity between the sexes. The line of argument that I have pursued provides a way of measuring the magnitude of the imbalance. I have broken down the sharing process into two stages: sharing among producers, followed by distribution to the rest of the group (the assumption here is that the producers derive some utility from this distribution). In reality, the sharing takes place as a single process.

These people do not live a hand-to-mouth existence. They may eat a little of what they obtain in the field, but most is transported to the camp and shared out. Therefore the earnings are purely notional, in the sense that their value is dependent upon an exchange process. If there are many people in the camp, then the rate of exchange is low, if there are few, then the exchange rate is high. What a hunter or forager brings back from a particular day's work is of no direct benefit, because it is pooled with the produce from other foragers and shared out to all members of the group. It is as if the forager received a wage that is determined by the success of the company as a whole.

Whether foraging or doing camp chores, the individual is working for payment that is delayed. The payment comes in the form of shared-out food. If the payment had come in the form of money, it would have to be spent on that same shared-out food, because there is nothing else to buy. We can use these figures to calculate the notional wage earned by hunters and gatherers. In reality, the producers receive the same income as the non-producers (in relation to their requirements) and there would seem to be little room for individuality in this society. We should not assume, however, that the productive individual does not benefit over and above the non-productive. The San live in small groups and can easily remember individual characteristics. It may well be that the productive ones benefit at the important stages of their life-cycle, such as in choice of a mate.

CHAPTER FOUR

Sleep as an ethological problem

Sleep takes up about one third of the animal's time, and accounts for an even greater proportion of the life of the neonate (Meddis, 1983). Yet it is largely ignored by ethologists, and it is hard to understand why this should be so. That the neglect is due to ignorance can hardly be the case since the seminal paper on sleep in animals, by Meddis (1975) appeared in *Animal Behaviour*. Perhaps ethologists do not think of sleep as real behaviour. Perhaps sleep is simply considered to be unimportant. Perhaps ethologists think that the problems of sleep have already been solved by other scientists. My aim is to show that sleep is an ethological problem.

Diagnosis of Sleep in the Animal Kingdom

Ostensibly, sleep is not uncommon. Many animals show periods of apparent sleep, rest or torpor. For example, the Mediterranian flour moth *Anagasta kuehuiella* appears to adopt a special sleep posture. The antennae are usually directed forward during locomotion, but during apparent sleep they are laid backwards and may be crossed over so that their tips are hidden beneath the wings (Andersen, 1968). When in this posture a wing can be lifted with a fine brush and allowed to fall back, without arousing the moth. Normally the slightest touch would cause the moth to fly away. During these quiescent periods the animal is resistant to disturbance, but is it really sleeping? Similarly, a chameleon settles on a branch before sunset and curls up its tail. It becomes immobile and will not attack insects even when they alight on its body. Shortly after sunset the eyes close and the animal appears to be asleep. It will usually remain in this position throughout the night (Meddis, 1977).

Are these animals really sleeping? We can answer this question only if we have agreed criteria of sleep. One commonly used criterion is the arousal threshold. A stronger stimulus is required to arouse a sleeping animal than an awake animal. For example, Amlaner and

McFarland (1981) induced herring gulls to rest on an electric grid in their normal territorial habitat. Starting from various initial postures, the experiments increased the electric current from zero to an amount just sufficient to cause the bird to change its posture. They found that when the bird was in a typical alert posture or typical rest posture a low current was required to induce a change in posture, but when it was in a typical sleep posture with slowly blinking eyes, a higher current was required. By this arousal criterion, the flour moth, the chameleon and the herring gull would all be judged to show true sleep behaviour.

Another criterion has been developed from studies of sleep in humans. Measurements of the gross electrical activity of the brain produce an electroencephalograph (EEG). This method was pioneered by Berger (1929), who discovered a typical rhythmic EEG activity in the resting, relaxed, person. Berger called this the alpha rhythm. He found that the rhythm, which has a frequency of about 8–12 Hz, is interrupted when the subject opens his eyes and pays attention to something. Subsequent research showed that the 'Berger rhythms' change markedly during sleep. During the early stage of sleep, periods of low amplitude, high frequency (14–16 Hz) rhythms are found. In deeper sleep, large amplitude, low frequency (3–4 Hz) delta waves are seen. However, during deep sleep there are also periods of low amplitude, fast EEG activity. This appears to be paradoxical because the pattern is typical of an awake person, although the subject is really in a deep sleep. During such paradoxical sleep, rapid eyeball movements (REM) often occur, and if the subject is awakened he reports that he has been dreaming. No dreams are reported when the subject is awakened at other times. Dreams occur periodically at about 90-min intervals, giving rise to five or six dreaming sessions per night. When studied continuously throughout their sleeping time, some other animals show similar paradoxical sleep episodes in their EEG. Modern nomenclature recognizes two main types of sleep in mammals and birds: quiet sleep with synchronized EEG without obvious signs of dreaming, and active sleep with desynchronized EEG and dreams reported by humans.

At first, it was thought that dreams did not occur during quiet sleep. However, it is now known that dream reports do occur when people are woken during quiet sleep, although the dreams seem to be less vivid and story-like. This is not very surprising, because the brain is less active during quiet sleep. There is little evidence that dreams are necessary for our mental well-being, or that their content has any special relevance to our waking life. However, it is more likely that the mental activity which occurs during active sleep is more vivid and more likely to be remembered (Foulkes, 1983). Many sleeping animals show REM that are accompanied by physiological activity similar to that found in human beings. We cannot wake animals and ask them about their dreams, but we can detect many of the phenomena that accompany human dreams, including bodily movements. When we observe movements, or vocalizations, in a sleeping cat or dog we usually assume that they are dreaming. On the basis of the evidence, many workers are inclined to agree that some animals probably experience dreams that are akin to those of human beings, although this interpretation has recently been questioned (see below).

Many physiologists define sleep in terms of the EEG pattern. In mammals this criterion correlates well with the arousal criterion, but this is not the case with all animal groups. The EEG patterns of birds are very similar to those of mammals, but the EEG patterns found in reptiles bear no resemblance to those of birds or mammals, and the gross electrical activity of the other vertebrates is even more remote from the mammalian pattern. Thus the EEG criteria of sleep apply only to birds and mammals, which constitute a very small part of the animal kingdom. Some physiologists go so far as to conclude that only birds and mammals show true sleep. Behaviourally, however, it would appear that sleep does occur in fish, amphibians and reptiles. Many fish become inactive during the hours of darkness, and they are often slow to react to disturbances. Some can be handled for a minute or so, before they become aroused and swim off. Crocodiles spend long periods in a state of complete immobility. Sometimes they are waiting for prey, but at other times they can be approached and even picked up. From a biological point of view, these animals would appear to be asleep, even though they do not show the typical EEG pattern of birds and mammals.

Sleep in Birds and Mammals

Sleep in birds has tended to be a relatively neglected topic. In a survey by Amlaner and Ball (1983) some 70 species are tabled in connection with sleep posture, and about 30 for the temporal characteristics of sleep of birds in the wild. Much of these data are incidental, and not terribly reliable. More recently, a more concerted effort has been made to get to grips with problems associated with sleep in birds. This work was started in Oxford by Amlaner, Ball, Lendrem and McFarland, and has now shifted to the USA (Amlaner and Ball, 1988).

The general characteristics of avian sleep are very similar to those of mammals (see below). Avian sleep is characterized by periods of inactivity, which may be as long as 13.4 hours or as short as 0.08 hours. Birds sleep in species-typical sites, usually well protected from predators. They may take up one of a number of typical sleep postures, as described by Amlaner and Ball (1988). During sleep there is a raised threshold of arousal, and a typical EEG pattern. However, the picture is complicated by the number of types of sleep that have been described.

Amlaner and Ball (1988) describe the EEG characteristics of sleep proper, comprising quiet sleep (QS) and active sleep (AS), and various intermediate arousal states, such as drowsiness, cataleptic sleep (alert resting), and unihemispheric quiet sleep. In the last of these, the bird has one eye closed, while the contralateral hemisphere shows a QS-like EEG pattern, and the ipsilateral hemisphere shows a pattern resembling wakefulness. The metabolic rate during unihemispheric QS is similar to that of the bird during bilateral quiet sleep.

Mammalian sleep has a number of general characteristics in common with that of birds. It involves prolonged periods of inactivity organized on a rhythmic daily basis. The animal usually seeks out a typical sleep site and adopts a typical sleep posture. During sleep there is a raised threshold of arousal. The EEG shows high-voltage, slow waves during quiet sleep and low voltage, irregular waves during active (REM) sleep. However, if we look at these characteristics in turn, we see that there is considerable variation among species (Meddis, 1983).

The daily periods of inactivity may vary from about 20 hours in the oppossum to about 2 hours in the giraffe. The human average is 8 hours, but this may vary between individuals. Some people habitually sleep 16 hours a day, while others sleep only 1 or 2 hours. A daily sleep ration as low as 45 minutes has been reliably recorded (Meddis, 1977). People who habitually sleep for a very short period each night appear to be perfectly normal in other respects, and this finding has cast doubt on the traditional view that sleep is physiologically necessary, or that lack of sleep is harmful.

Functions of Sleep

Traditionally, it has been assumed that sleep serves some restorative function. Aristotle noted that 'if (sensory perception) continues beyond its appointed time limit . . . it will lose its power and will do its work no longer'. 'Sleep . . . is a state of powerlessness arising from excess of waking . . . and rising of necessity . . . for the sake of conservation.' He also held 'that the nutrient part does its work better when the animal is asleep than when it is awake. Nutrition and growth are then especially promoted'. In other words, wakefulness, and its associated activities, results in a state (tiredness), which is antithetical to efficient activity. Tiredness promotes a state of sleep that is necessary for restoring behavioural efficiency, and is beneficial in promoting metabolism and growth.

Modern versions of this theory vary in their interpretation of the restorative process. Some regard sleep as a time for recuperation and repair of the body and the brain, others as a time for disposing of accumulated toxins, etc. An implication of most restorative theories, however, is that the recuperative process is necessary for good health, and that the animal would suffer serious deleterious consequences without it.

It has also been suggested that dreaming is necessary for our mental health. However, this interpretation has recently been questioned (Foulkes, 1983). To understand this challenge we have to look at some of the characteristics of human dreams, as reported by people woken during the different phases of the sleep cycle. Firstly, the dreams that occur in active sleep usually involve visual imagery, unlike those of quiet sleep, which are often reported as being verbal or imageless. Secondly, the dreams experienced during active sleep are 'manifest' dreams which involve constructive rearrangement of past experience. Dreams that occur in quiet sleep are relatively unstructured. Thirdly, the manifest dream is a narrative and not a succession of unrelated memories, such as occurs during quiet sleep. Fourthly, the manifest dream is an hallucination in that we do not question the truth of the story whilst we are dreaming. Some dreams that occur during quiet sleep may involve some critical judgement and are more akin to ordinary thinking.

It has been suggested (Foulkes, 1983) that the manifest dreams that occur during active sleep involve cognition of a peculiarly human sort. This view rests upon the assumption that other animals are not capable of the linguistic type of cognition required for manifest dreams. In the case of some animals, such as chimpanzees, this is an issue upon which the behavioural scientists are divided (Foulkes, 1983). Nevertheless, it is difficult to argue with the idea that there are some types of human dream the like of which does not occur in other animals. We cannot imagine a dog dreaming that it was taking part in a debate. The evidence from human dream research does suggest, therefore, that the type of dream that probably occurs only in humans is more likely to be reported upon waking from active sleep than from quiet sleep. An implication of this conclusion is that animals which appear from their behaviour to be dreaming during active (REM) sleep are probably not dreaming in the way that humans do. It is therefore not very sensible to use human dream content as a guide to a functional theory of dreaming, or of sleep.

That animals are motivated to sleep, from time to time, is beyond dispute. The question is do animals need to sleep in the sense that they will suffer some kind of physiological damage if they do not sleep? Thus the essence of a restorative theory is a functional one, and not a matter of the motivational mechanisms that induce animals to sleep. In the case of feeding behaviour, we can say that animals need to feed, because they will suffer physiological damage if they do not do so. We can also say that there are appropriate motivational mechanisms that ensure that the animal seeks food when it needs food. In the case of sexual behaviour, we cannot say that the individual animal needs sex to survive, but we can say that animals are strongly motivated towards sexual behaviour from time to time. Restorative theories of sleep claim that sleep is akin to feeding in these respects. Non-restorative theories claim that sleep is more like sex than feeding.

Supporting evidence for restorative theories of sleep is weak. Despite years of research, there is no hard evidence as to the nature of the vital restorative processes that are supposed to occur during sleep. Various processes have been suggested. For example, that sleep is necessary for memory consolidation (Fishbein, 1981); for learning (Jouvet, 1978, 1980); for tissue repair and growth (Adam and Oswald, 1977; Adam, 1980); for body restitution, etc. In none of these cases does the evidence stand up to scrutiny (Horne, 1983; Webb, 1983).

We now come to the question of possible alternative theories. Many have been proposed in recent years. Thus Allison and Van Twyver (1970) suggested that the sleep characteristics of each species was evolutionarily determined by the predator-prey relationships, and that animals slept in a safe place as a protection against predators. Meddis (1975) proposed that sleep serves to immobilize the animal during periods during which it would otherwise be dangerous, wasteful, or counter-productive for the animal to be active. Webb (1975) maintains that 'periods of non-responding are necessary for survival and sleep serves as a necessary condition to aid in maintaining these periods of non-responding'. We may designate these theories as protective theories, in which sleep has evolved as a means of keeping the animal out of trouble at those times of day when it would otherwise be vulnerable.

Assessment

In assessing theories of sleep function we have two main questions to take into consideration. Firstly, is there any theory which can adequately account for the known facts about sleep? Secondly, is the quality of the evidence sufficient to make the theory credible?

There is no doubt that a restorative theory could, in principle, account for the main features of sleep in animals. The problem with restorative theories is the poor quality of the evidence. Non-restorative theories are not able to provide a convincing explanation of certain aspects of sleep, and they also suffer from poor evidence, though perhaps not so seriously as do the restorative theories.

It has been claimed by some workers that differences in biochemical activity, between the sleeping and waking states, provide evidence for restorative theories. These include differences in hormonal activity, protein turnover, and rate of mitosis. In a recent review, however, Horne (1983) casts doubt on this type of evidence. He notes that many of the claims have proved difficult to substantiate, and some of the evidence is contradictory. Moreover, even if there are some genuine biochemical differences, they may have nothing to do with sleep function, but may merely reflect the existence of the sleep-inducing mechanism. All agree that there must be some mechanism that induces sleep, which must manifest itself in some aspects of the animal's biochemical machinery. Similar arguments apply to those aspects of the neurophysiological and neurochemical activity of the brain which seem to occur only during sleep (Horne, 1983). To show that such activity forms part of a vital restorative process it would be necessary to demonstrate the functional significance of the process, and not merely that it is associated with sleep.

One approach to this problem has been to deprive animals of sleep and look for behavioural and physiological changes during deprivation and recovery. It has been claimed that humans deprived of sleep become irritable, and even irrational, and perform badly in various psychological tests (McGrath and Cohen, 1978). There have also been many claims involving physiological changes (reviewed by Horne, 1978). It has even been claimed that sleep-deprived rats die sooner than controls (Rechtschaffen et al., 1983). In addition, it has been claimed that sleep deprivation results in various rebound effects during recovery. Especially significant is the claim that animals catch up on the lost sleep. This is seen by some as strong evidence in favour of restorative theories. However, it is evident that animals do not catch up on all the lost sleep (Horne, 1983), and the rebound effect occurs only under some circumstances. Moreover, as we see below, a rebound effect does not necessarily imply that a restorative function is involved. Sleep-deprivation studies have also been criticized for employing unnatural, and even traumatic, methods of keeping the animal awake. A fairly common experimental practice is to maintain the animal in an apparatus in which loss of muscle tone results in the animal being automatically tipped into water (Horne, 1983).

Sleep-deprivation studies are also complicated by circadian rhythms. The motivation to sleep follows a distinct daily rhythm in many animals, including humans. In a large and systematic study, Czeisler et al. (1980) found that there were no significant correlations between observed sleep duration and length of prior wakefulness in humans. Instead they found that circadian peaks of tiredness corresponded with fluctuations in body temperature. When self-selected bedtimes coincided with a temperature trough, the average sleep duration was 7.8 hours, whereas if a person chose to go to bed at the temperature cycle maximum, sleep duration averaged 14.4 hours. In general, individuals develop the habit of sleeping at particular times, and when deprived of sleep they may feel tired when the normal sleep time occurs, but they may feel less tired when this time has passed. The results of sleep deprivation studies, therefore, depend critically on the relationship between the deprivation period and the normal sleep cycle.

In general, sleep-deprivation studies show that partial deprivation has no effect on other aspects of behaviour, but chronic deprivation may have an effect. However, it is not known whether the apparently deleterious effects of sleep deprivation are due to lack of some restorative function, or are merely motivational. In particular, it has not been demonstrated that disruption of the normal sleep cycle has a more disruptive effect than disruption of other behavioural cycles. On the whole, sleep-deprivation studies have been poorly designed, and have failed to control for the strong motivational components of circadian rhythms, or for possible stressful effects of unnatural methods of preventing sleep.

Non-restorative theories of sleep recognize that animals are motivated to sleep. Many of the physiological and behavioural differences between the awake and the sleeping animal can be attributed to the motivational mechanisms responsible for inducing sleep. Non-restorative theories, therefore, have not been particularly concerned with the biochemical details attending sleep, sleep deprivation and recovery. An ethologist would not be surprised to find some sort of rebound phenomenon following sleep deprivation, because such rebound occurs after deprivation of a wide variety of activities, including feeding, courtship display, nest care, etc.

To the ethologist, rebound phenomena are an indication that certain motivational mechanisms are at work. In some cases these mechanisms may involve a build-up of motivational potential, but in other cases they may not. Even if rebound phenomena discovered from sleep deprivation studies proved to be due to some motivational build-up, the implications for functional theories of sleep would not be profound. If it is important (for some unknown reason) for animals to sleep at certain times, then it would not be all that surprising to find that there was a build up of motivational potential for sleep at that time, and a motivational rebound resulted from prevention of sleep at that time. Analogous phenomena occur in the motivational control of many other aspects of behaviour, including feeding, sexual and parental behaviour (see Chapter 2). The motivational mechanisms employed to achieve a particular evolutionary end are not always the obvious ones, and ethologists have learned not to draw too hasty conclusions about biological function from motivational or physiological phenomena.

What then of positive evidence for non-restorative theories of sleep? Various types of statistical argument have been deployed. These usually involve comparison of species. Thus Hartmann (1968), using data from six mammalian species, reported a strong inverse correlation between the length of the quiet/active sleep cycle and the average metabolic rate for the species. Zepelin and Rechtschaffen (1970, 1974) found this relationship to be valid for a further 18 species, but they also found that similar relationships exist between various measures of sleep and a number of other variables, including life-span, body weight, brain weight, and gestation period. Based on correlations with ecological indices, Allison and Van Twyver (1970) claim that the status of a species as a predator or prey and the security of the typical sleeping arrangements are the decisive factors with respect to sleep time. They found that species with short sleep times are those that sleep in exposed places, whereas animals with long sleep times are either predators, or sleep in very safe places. Allison and Cicchetti (1976) devised five-point scales for predation, sleep exposure and overall danger. The predation index was their assessment of the extent to which a species is preyed upon in its natural habitat. Their sleep-exposure scale represents the degree of protection offered by the sleep site typical of the species. Their scale of overall danger is a general estimate of danger from predators, taking both predation level and sleep exposure into account. They found that all these measures were correlated with sleep parameters.

A reanalysis of all available data is offered by Meddis (1983). His main finding is that the amount of active sleep in adult animals can be accounted for in terms of exposure to predatory threat and the degree of neonatal brain advancement that is characteristic of each species. Species exposed to predatory danger have less active sleep than average, while species whose major growth of nervous tissue occurs after birth have more than average active sleep as adults. Smaller species and species typically exposed to danger, have shorter episodes of active sleep. Grazing animals have short sleep times, and because they are typically large animals this gives the impression that there is a relationship between body size and amount of quiet sleep. However, Meddis (1983) claims that this is a statistical artefact and that there is no such relationship among non-grazing animals.

A problem with much of the comparative research on sleep is that it suffers from a number of methodological shortcomings. Firstly, the species is generally used as the unit of statistical analysis. Species can rarely be treated as independent points in comparative analysis, because species within taxa may share characters as a result of common ancestry rather than convergent evolution. A statistical analysis of species points can yield a significant but spurious relationship (Elgar and Harvey, 1987; Pagel and Harvey, 1988). Comparative studies can also suffer from the effects of confounding variables (Clutton-Brock and Harvey, 1984; Pagel and Harvey, 1988). As we have seen, Meddis (1983) disputes the negative correlation between body weight and total sleep time claimed by Zepelin and Rechtschaffen (1974) and Allison and Cicchetti (1976). Meddis (1983) claims that the correlation is influenced by the fact that grazing animals, which are typically large, have short sleep times. When Meddis removed the large grazing animals from his analysis, the correlation became non-significant.

Paul Harvey's research group, which specializes in comparative analysis is well placed to rectify these shortcomings. They have reanalysed the available data on 69 species of mammals, belonging to 13 orders. To determine the appropriate taxonomic level of analysis, they employed nested analysis of variance, and chose the family as their unit of analysis (Elgar, Pagel and Harvey 1988). They found that TST (total sleep time) and QST (quiet sleep time) are strongly negatively correlated with adult body weight, but the correlation of AST (active sleep time) with body weight was not significant. Whereas Meddis (1983) found no relationship between QST and body weight after removing large grazing animals from his sample, Elgar, Pagel and Harvey (1988) found that body weight is significantly negatively correlated with both TST and QST. This difference is probably due to the classification of grazing animals. While Meddis (1983) included mainly artiodactyls, Harvey's group included a wider taxanomic range of grazers.

The relationship between body size and active sleep is of considerable interest because of recent evidence that thermoregulatory responses, such as sweating, panting, shivering, and peripheral blood flow changes, are disrupted during active sleep (Parmeggiani, 1977; Walker et al., 1977; Toutain et al., 1972). These thermoregulatory mechanisms function normally during quiet sleep, but give no response to thermal stress during active sleep. Such lack of thermoregulatory response is much more serious for small animals whose body temperature is more easily influenced by ambient temperature than large animals with greater thermal capacity. It is of interest, therefore, to find that small animals have shorter episodes of active sleep than large animals. It is possible that each period of active sleep is terminated before it becomes thermally risky. Support for this idea comes from studies of hibernation in mammals and torpor in birds, which show that such states are associated with large reductions in the proportion of active sleep (Walker and Berger, 1980).

Hibernation requires continuous thermoregulation to prevent body temperature falling to lethal levels. The lower the body temperature the less the proportion of active sleep, and in some species active sleep is reduced to zero. Thus a possible function of quiet sleep is to provide for long periods of immobility while maintaining normal thermoregulation. It has even been suggested that the purpose of active sleep is to save the energy normally expended in thermoregulation (Horne, 1977; Walker and Berger, 1980).

Paul Harvey's group found that AS–QS cycle length has very strong positive correlations with both adult body weight and basal metabolic rate (Elgar, Pagel and Harvey, 1988). Thus smaller mammals have shorter active sleep episodes, a result that agrees with those of Zepelin and Rechtschaffen (1974) and Meddis (1983). However, they also found that the correlation of AS–QS cycle with metabolic rate is not significant after controlling for the effects of body size. In other words, mammals with high energy expenditure for their size do not have shorter active sleep episodes. This is contrary to what would be predicted by the hypothesis that animals have evolved active sleep to reduce heat loss.

The main problem for non-recuperative theories of sleep is that they fail to account for many of the characteristics of sleep. If the prime function of sleep is to enable the animal to be immobile at appropriate times and so avoid predators and save energy, then why is sleep accompanied by raised thresholds of arousal? Immobility with vigilance is achieved during drowsiness in ungulates and carnivores (Meddis, 1983). However, this drowsiness would not appear to be a substitute for sleep, because cattle sleep little whereas cats and dogs sleep a lot.

Non-recuperative theories also have difficulty in explaining why there are different types of sleep. Quiet sleep and active sleep are very different from each other. Indeed they have little in common apart from the fact that they occur while the animal is in the same place, and while it has a raised threshold of arousal. Non-recuperative theories can offer little explanation of the quiet/active sleep cycle, apart from postulating that it has something to do with thermoregulation. Meddis (1977) argues that active sleep is similar to reptilian sleep and is vestigial in warm blooded animals, having no function. In this way, Meddis can account for the lack of thermoregulation during active sleep. However, it is difficult to see how such an apparently disadvantageous evolutionary vestige could have survived for so long and be so widespread among birds and mammals.

What can we conclude about the nature of sleep? We know that many species have periods of dormancy during which they show raised thresholds of arousal. Only warm blooded animals, however, show true sleep with characteristic EEG patterning. The remarkable similarity in the sleep pattern of birds and mammals suggests that sleep is connected with thermoregulation, but it seems unlikely that this is its primary function.

The traditional idea that sleep serves some vital restorative function does not seem to be supported by the evidence, and leaves us with the problem of explaining why some species sleep very little. The evidence suggests that sleep is controlled by motivational mechanisms based partly on circadian rhythms and partly on regulatory principles (Daan, 1982). The nature of the motivational organization, however, tells us little about the function of sleep. Non-recuperative theories of sleep function can account for the variety of sleep patterns found among animals, but do not provide a satisfactory explanation of the intrinsic phenomena of sleep. In particular, there is growing evidence (see below) that animals incur considerable costs by sleeping, and this implies that they must reap some benefit. Of special interest here is the evidence that some aquatic mammals (Mukhametov and Poly-akova, 1981), and some birds (Amlaner and Ball, 1988), have uni-hemispheric sleep. In other words, part of the brain sleeps while the rest remains awake. It appears that normal (bihemispheric) sleep in these species would be deleterious, because of breathing difficulties (in aquatic mammals) or vigilance problems (in birds). If this view could be substantiated, then the implication would be that there is some evolutionary difficulty in solving these problems by doing without sleep. Remember, however, that some species sleep very little.

Some scientists favour a dual theory, and distinguish between obligatory sleep with serves some restorative function, and facultative sleep which fulfils an ecological role (Horne, 1983). This type of theory can account for many of the known properties of sleep, but still leaves unexplained what it is that is restored during obligatory sleep. The problem with sleep research is that it has been neglected by ethologists and has not been integrated with other aspects of behaviour. Sleep should be studied, not in isolation, but as part of the animal's total behaviour repertoire.

The Problem for Ethologists

Sleep poses a number of problems for biological scientists in general, but two problems are primarily ethological. The first is mainly empirical: when, and under what circumstances do animals sleep in the wild? The vast majority of sleep studies have been conducted in the laboratory, where it is easy to diagnose sleep, but where it is difficult to conclude much about the normal circumstances of sleep. In the field, it is difficult to diagnose sleep (see above), and it has too often been inadequately identified. The circumstances in which animals sleep has sometimes been described, but (as we see below) this should really be done in relation to the animal's whole behaviour repertoire.

The second problem concerns the function of sleep. As we have seen, many scientists have sought to identify the function of sleep in the brain, and its role in the organization of behaviour. Few have addressed the question of function in the more evolutionary sense. It is here that ethology has an important role to play. In assessing the functional significance of any behaviour we need to know what the consequences of the behaviour are for the animal, in two differing respects. We need to know (i) how the behaviour alters the animals own state; and (ii) how the behaviour affects the animal's relationships with its environment (i.e. the costs and benefits of sleep). Let us consider each in turn.

There are a number of ways in which sleep could affect the animal's physiological state:

(a) There will be direct effects of sleep on the body resulting from the sleeping situation and the sleep posture. For example, if the animal chooses a sheltered place to sleep, there will be consequent changes in the thermal and water relations with the environment.

For example, glaucous-winged gulls conserve about 15 per cent of the average daily energy expenditure by sleeping rather than resting (Opp and Ball, 1985). Three simultaneous mechanisms contribute to this energy saving: reduced heat production, lowered body temperature and decreased thermal conductivity (Opp and Ball, 1987).

(b) There may be direct effects of brain activity upon the physiological state. Thus there may be induced changes in metabolic rate, and changes in endocrine activity. Of particular interest here is the question of whether these changes are beneficial in themselves, as would be implied by a recuperative theory of sleep (see above).

(c) There will be indirect changes resulting from the fact that the animal is not doing other kinds of behaviour. It is not eating, so its hunger will be gradually increasing. It is not active, so its energy expenditure will be reduced.

Thus the consequences of sleep on any particular occasion will depend upon the animal's physiological state at the time, the consequences of sleeping in a particular place (the effects of the microclimate, etc.), and the behaviour that the animal would have been doing if it were not sleeping (i.e. the postponed behaviour).

We now have to consider the costs, benefits and behaviour which results from the animal being in a particular state. On the basis of evolutionary theory may we come to the conclusion that the animal is designed by natural selection to take whatever action minimizes the cost under the circumstances (see Chapter 1). The (instantaneous) cost is the decrement in Darwinian fitness that would occur if no remedial action were taken. The animal will have various physiological and behavioural options, subject to certain constraints. The task for the ethologist is to specify these options, and to (i) determine the mechanisms by which the individual animal decides which option to take;

and (ii) specify the course of action that would minimize the cost in a given situation. The first of these tasks is essentially a motivational problem and has been discussed in Chapter 1. The principles governing the motivation of sleep behaviour cannot be different from those governing other types of motivation. To tackle the second it is necessary to consider both the cost of being in a particular state, and the cost of taking particular remedial action.

In general, the cost of being in a particular state is an increasing function of the deviation of the state from the optimal state. Thus if the optimum body temperature for a mammal is 37 °C, temperatures above and below this carry increasing cost. The cost is incurred partly as a result of reduced physiological efficiency, but primarily because of the increased risk of crossing a lethal boundary. For example, a hungry animal may not be in immediate danger of death from starvation, but there is always a risk of a future lost opportunity to obtain food. The hungry animal may not be able to obtain sufficient energy when needed, because food has become temporarily unavailable, or because of the presence of predators, or because of a change in food quality. In the case of some state variables, such as those relating to reproductive state, there are no obvious lethal boundaries. The cost may, nevertheless, increase with deviation from the optimal state, because there is an associated decrement in fitness. (An animal that is not in the optimal reproductive state at the right time is likely to have reduced fitness). A third possibility is that there are some state variables for which deviation from the optimum bears very little cost, implying that the associated behaviour is of little functional importance. In the case of sleep, we do not really know what the cost functions are like. An extreme form of a deprivation theory would imply lethal boundaries on some variables related to sleep. At the other extreme, the theory that sleep is merely a time-filler would imply relatively little change in the cost associated with sleep variables.

In theory, the animal chooses its behaviour so as to change its state in a way that provides minimum cost (or maximum benefit in terms of fitness). This is a type of optimal control problem of a type which has received considerable attention in recent years. The solution depends upon knowledge of the animal's state, its behavioural options, and its cost function (the cost function specifies the relations among cost, state and behaviour). We assume that, in general, the animal is designed to change from one activity to another when the net cost of the latter becomes lower than that of the former. However, there may also be costs involved in the activity of changing itself, as discussed in Chapter 1.

In the case of sleep, we must assume either that the animal sleeps when the benefits of sleep outweigh those of other activities, or that a strict cost-benefit analysis cannot be applied to sleep. There are two basic ways in which the benefits of sleep could become greater than those of all other activities:

(a)   When the animal is not sleeping, there is a build-up of a need to sleep, which carries associated costs. The benefit gained from sleeping is that these costs are reduced:

(b)   At the time that the animal normally sleeps, the benefits to be gained from other activities are reduced, because of changes in the environment. For example, some animals cannot forage efficiently in low illumination (e.g. Kacelnic, 1979). To forage at night would incur increased risk from predators (high cost) and reduced profitability (low benefit), compared with foraging during the day. Similar arguments may apply to other activities, leaving sleep as the only activity worth doing.

Of course, it is entirely possible that the true situation is a mixture of these basic possibilities. It is also possible that cost-benefit analysis does not apply to any particular situation, but that the animal sleeps according to a preprogrammed routine, which provides the greater benefit on average. Nevertheless, it is still worth enquiring what the costs and benefits of sleep might be, because the mechanism responsible for the sleep routine must, at some stage, have been subject to natural selection.

Let us first consider the possible costs and benefits attributable to sleeping itself. On the benefit side, there may or may not be some biophysical, neurological, or physiological effect on the brain that is beneficial, or possibly even essential for life. The evidence, as we have seen above, is not very convincing despite years of research on this aspect of sleep. There may also be benefits resulting from the muscular relaxation that occurs during sleep, and there is almost certainly some benefit to be gained from the lower rate of energy expenditure that is characteristic of sleep. Finally, in some species, there is possibly some antipredator benefit to be gained from the choice of a safe sleep site.

On the cost side, there is very likely to be some cost associated with the loss of vigilance that occurs during sleep. Evidence for this comes from a number of studies (Lendrem, 1983, 1984; Ball, Shaffery and Amlaner, 1984). Even among species that are not themselves preyed upon, loss of vigilance may have deleterious social effects, such if a sexual or political rival can take advantage of the situation (de Waal, 1982).

There are also possible indirect costs resulting from the postponed behaviour. Thus the sleeping animal may have increasing hunger, may be taken advantage of by a rival, may miss a reproductive opportunity, etc. This aspect of the cost of sleeping is complicated by the phenomenon of behavioural resilience. For example, if an animal's normal daily food intake was less than normal, we might expect it to spend less time sleeping and more time foraging. It may be, however, that the animal would not spend less time sleeping under any circumstances. In other words, the system controlling sleep will be somewhat resistent to competition from other motivational systems. The extent to which an activity is resistent to pressure from other activities is called behavioural resilience.

Theoretically, behavioural resilience is a design feature that can be derived from the cost function mentioned above (Houston and McFarland, 1980; McFarland and Houston, 1981). It can be measured, either by manipulating the animal's total daily behaviour routine, or by determining demand functions. These aspects of behavioural resilience are discussed in Chapter 2. For our present purposes it is sufficient that the notion of behavioural resilience provides a measure of the long-term importance of an activity (more correctly, the consequences of the activity) to the animal. For example, if sleep fulfills a vital function, we would expect the behavioural resilience of sleep to be high. When the animal has insufficient time to perform its daily quota of foraging, territorial defence, or other behaviour, we would expect the time spent on sleep to be reduced only slightly, at the expense of other less resilient activities. On the other hand, if the consequences of sleep are not very important, then we might expect sleep to have low resilience. When short of time, the animal will cut down on the amount of sleep so as to make time available for other activities with higher resilience. If, as has been suggested, (Meddis, 1975) sleep is merely a time-filler, it should be possible for the animal to give up sleep entirely when it has more important things to do. Sleep could then be classed as a leisure activity (McFarland, 1985).

In the final analysis, what matters in considering the function of sleep is (i) the change in fitness that would be incurred if the animal slept more or less than it does normally; (ii) the change in fitness that would result from sleeping in a different place, or at a different time from normal, and (iii) the change in fitness that would occur if the animal slept in a different manner. The ethologist would have to have some idea about the answers to these questions before attempting to integrate sleep into a model integrating the animal's behaviour repertoire and life-style. It is, perhaps, not so puzzling that animals sleep, given what we know about energy conservation etc., but that they sleep in such a peculiar manner. Why do sleeping animals shut down sensory input, alternate between active and passive sleep, lose muscle tone, and thereby risk predation?

CHAPTER FIVE
Goal-directed behaviour

Many aspects of animal behaviour are goal-achieving. These include various aspects of homeostatis, such as body-weight regulation and thermoregulation, searching and homing behaviour, feeding and drinking, nest-building, and many more complex aspects of motivation and learning. Conventional wisdom has it that many of these are not only goal achieving but goal-directed. That is, the explanation of the goal-achieving behaviour includes an internal representation of the goal-to-be-achieved which is instrumental in guiding the behaviour. My purpose here is to question the conventional wisdom. I will question the nature of the evidence that is generally put forward in support of the goal-directed view. I will suggest that there are alternative explanations that adequately account for the behaviour without involving the concept of goal-directedness, and I will argue that the goal-directed model cannot account for complex, real-life behaviour.

Terminology

A system can be goal-achieving or goal-seeking without being goal-directed. A goal-achieving system is one which can recognize the goal once it is arrived at (or at least change its behaviour when it reaches the goal), but the process of arriving at the goal is largely determined by the environmental circumstances. An example is epigenesis in which each stage of ontogenetic development sets the stage for, but does not dictate the next. What happens at each point in the causal chain depends partly on what has just happened, and partly on the environmental circumstances. A similar principle can be seen in immunological and other biochemical systems, in which there is some form of template recognition. In such systems the significant events occur when one molecule recognizes another (a kind of lock-and-key mechanism) within a suitable environment, or medium. These systems are goal-achieving by virtue of the fact that matters are so arranged that the necessary environmental features are generally present at the appropriate stage in the causal chain. For example, newly hatched ducklings tend to follow whatever moving object they see. In nature, matters are so arranged that the moving object is invariably the mother. Bertrand Russell (1921) put forward this type of theory as a general theory of purposive behaviour: 'A hungry animal is restless until it finds food: then it becomes quiescent. The thing which will bring a restless condition to an end is said to be what is desired [its purpose]' (p. 32). The implication here is that the animal is in an environment in which the restless behaviour is appropriate. When it encounters, and recognizes, the relevant (food) stimuli, the animal changes its behaviour. This simple goal-achieving view of animal behaviour does not, in fact, accord with modern knowledge of foraging behaviour.

A goal-seeking system is one which is designed to seek the goal without the goal being explicitly represented within the system. Many physical systems are goal-seeking in this sense. For example, a marble rolling around a bowl will always come to rest in the same place. It may take various different routes, depending upon the starting conditions. The marble appears to be goal-seeking, because the forces acting on it are so arranged that the marble 'is pulled' towards the goal. In some goal-seeking systems that are designed to maintain a particular level, or direction, a dynamic equilibrium is maintained by self-balancing forces. Such systems may be based on osmosis, gyroscopic forces etc. The goal-seeking system achieves its effects by virtue of the forces acting on and within the system. There is no internal representation of the goal-to-be-achieved, nor does the system depend for its working on being in a particular environment, as does a goal-achieving system. Of course, the distinctions between goal-achieving, goal-seeking and goal-directed behaviour are somewhat blurred at the edges. I shall argue, however, that goal-directedness involves a difference of principle.

A goal-directed system involves an explicit representation of the goal-to-be-achieved, which is instrumental in directing the behaviour. In this essay I reserve the term 'goal-directed' to indicate behaviour (of a human, animal or machine) that is directed by reference to an internal representation of the goal-to-be-achieved. By 'directed', I mean that the behaviour is actively controlled by reference to the (internally represented) goal. The behaviour will be subject to outside disturbances, which will usually be corrected for. Thus by directed I mean that the behaviour is guided or steered towards the goal.

By 'goal-representation', I mean a physically (or physiologically) identifiable (in principle) representation that is 'explicit' in the sense of Dennett (1987, p. 216). In this category are included the 'set-point' of simple servomechanisms, the 'Sollwert', and 'search image' of classical ethology, and any form of explicit mental representation as postulated by cognitive ethology and some branches of psychology. Although, it can always be said of a goal-seeking system that the goal must be somehow represented, since there must be features of the system that are responsible for the goal-seeking behaviour, such representation is merely tacit. This tacit form of representation (Dennett, 1983) involves no more than naming the parameters of the system (which we normally think of as 'distributed' throughout the system), which may be said to 'represent' the goal, by virtue of their role in the design of the system. But this is true of any system which does what it does by virtue of its design, and it is not equivalent to the type of goal-representation which involves an explicit represen- tation that is capable of directing behaviour.

In conclusion, what do we mean by goal-directed behaviour? Since the goal of behaviour is achieved at the end of the relevant behaviour sequence, it is obvious that the goal itself cannot direct the behaviour. By goal-directed behaviour we mean that a representation of the goal directs the behaviour. If there is no representation of the goal, then the behaviour cannot be goal-directed. Thus this type of goal- achieving behaviour is dependent upon an explicit goal-representation.

Problems with Goal-Directed Behaviour

Ever since Rosenblueth, Weiner and Bigelow (1943) discovered that goal-achieving behaviour could be readily explained in terms of nega- tive feedback, this approach has become very fashionable. The problem is that theoreticians and model-makers, having found one way of solving the teleological problem, namely by postulating goal- representations, templates, cognitive maps and other forms of internal representation, have tended to apply it to all possible situations. Thus we see it in studies of animal motivation (e.g. Toates, 1986). In the guise of action theory, it has become the dominant theory in social and cognitive psychology (von Cranach and Harre, 1982; Frese and Sabini, 1985), and is prominent in artificial intelligence. The goal- directed model is commonly used where it is difficult to think of an alternative, but this does not mean that there are no alternatives. What started as an heuristic, became a band-wagon, and is now in danger of becoming a dogma.

To recapitulate, a system can be goal-achieving or goal-seeking without being goal-directed. A goal-achieving system is one which can recognize the goal once it is arrived at (or at least change its behaviour when it reaches the goal), but the process of arriving at the goal is largely determined by the environmental circumstances. A goal-seeking system is one which is designed to seek the goal without the goal being explicitly represented within the system. Goal-seeking systems can be based upon a dynamic equilibrium of various forces operating within the system. In engineering terminology these are sometimes called passive control systems (Milsum, 1966).

In the paradigm case of goal-directed behaviour, the difference between the 'desired' state of affairs and the actual state of affairs (as monitored in the negative feedback pathway) provides the (error) signal that actuates the behaviour-control mechanism, as shown in Figure 5.1. The behaviour may also be affected by disturbances, including influences from other systems. The exact nature of the input to the behaviour-control mechanism will vary from system to system. In some (called proportional control systems) the actuating signal is proportional to the error. In more sophisticated systems the control may be very complex. Engineers call this type of control system an active control system.

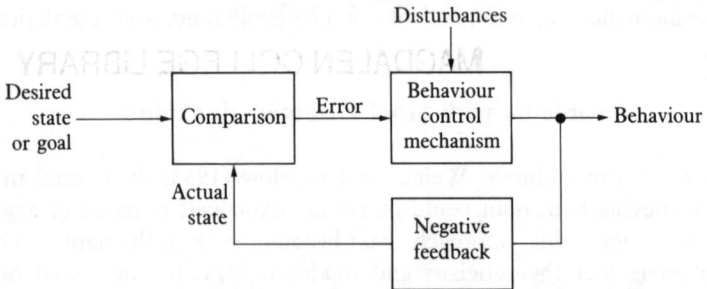

Fig. 5.1 The paradigm case of goal-directed behaviour. A comparison between the desired and actual state of affairs provides the (error) information that actuates the behaviour.

An essential feature of a goal-directed system is that some of the consequences of the behaviour are monitored and this information made available to the control mechanism. Few or many features of the behaviour consequences may be monitored, and the information gained may influence the control system in various ways. If, for some reason, it is not possible to monitor the behaviour consequences, then it is not possible to attain goal-directed behaviour. This may happen if the consequences of behaviour are not completely observable in the sense discussed in Chapter 6.

A simple example of a goal-directed system is provided by the thermostatic theory of temperature regulation in animals. True thermal homeostasis is found in birds and mammals, which are able to maintain a constant body temperature despite fluctuations in environmental temperature. They are called endotherms because their high metabolic rate provides an internal source of heat, and their insulated body surface prevents uncontrolled dissipation of this heat. (Animals that gain heat primarily from external sources like sunlight are called ectotherms.) Endotherms maintain a body temperature that is usually higher than that of their surroundings. The brain receives information about the temperature of the body and is able to exercise control over mechanisms of warming (such as shivering) and cooling (such as panting). According to the thermostatic theory, when the brain temperature gets too high (compared with some reference temperature), the cooling mechanisms are activated, and when it gets too low heat losses are reduced and warming mechanisms may be activated. The principle of operation is the same as that of a thermostatically controlled domestic heater.

The essential features of the thermostatic theory are:

(a) There is an internally represented reference temperature (often called the set point) with respect to which the body temperature is judged to be too low or too high.

(b) There is a mechanism (called a comparator) for comparing the set temperature with the body temperature.

(c) The output of the comparator actuates the heating and cooling mechanisms.

(d) The resulting body temperature is measured by appropriate sense organs (which we will call thermometers) and the comparison between set and body temperature is made on the basis of this information.

The overall principle of operation is called negative feedback, since the comparator essentially subtracts the measured body temperature from the set temperature, and the difference (called the error) provides the actuating signal, as illustrated in Figure 5.2. This is an active control system.

Passive control, or goal-seeking, systems have often been suggested

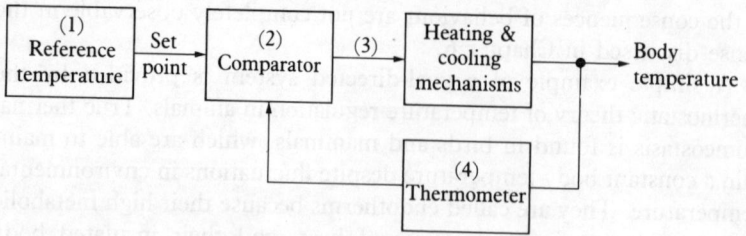

Fig. 5.2 The four essential features of a thermostatic theory. (1) the reference temperature, (2) the comparator, (3) the error signal, (4) measurement of body temperature.

as alternatives to the goal-directed explanation of physiological regulation. Thus for temperature regulation in mammals, in which the principle of operation is usually likened to that of a thermostatically controlled domestic heater, with a set-point representing the desired temperature, the alternative theory is that the thermoregulatory system is so designed that the processes controlling heating and cooling balance each other over a wide range of conditions, without any representation of a set-point. This alternative view has led some to argue that the set point concept has little more than descriptive value (McFarland, 1971; Mogenson and Calaresu, 1978).

In the case of thermoregulation the set-point concept seems intuitively satisfying but may be misleading. In other cases it may be counter-intuitive, but useful. For example, physiologists often refer to a set-point for the body weight of an animal. Many animals are able to maintain constant body weight with considerable precision and the set-point concept provides a useful means of portraying the regulatory system. It is difficult, on the other hand, to imagine that an animal has a device for telling it how heavy it should be, or that it has a means of measuring its own weight. Considerable controversy has arisen over the use of the set-point concept in this context (Wirtshafter and Davis, 1977).

It may be that the alternative theories of physiological regulation cannot be distinguished at the control-systems level of explanation. When the data can be interpreted in two different ways, it is possible that the alternative theories could be intertranslated. If this is the case, then only physical (or physiological) identification of the goal-representation (or set-point) could settle the issue. To my knowledge, this has not been done.

So far, I have argued that the goal achieved by an animal is often an emergent property of the system, and that it is a mistake to assume that the goal must necessarily have a representation in the animal, or in our model of the animal. This kind of argument can be applied at many levels, such as the level of peripheral control, of orientation, of physiological regulation or of motivation.

It is common practice to incorporate formal goal representations into models concerned with animal motivation. For example Toates (1986) writes as follows (p. 161):

> In the absence of stimuli associated with a primary incentive impinging upon the animal's senses, the animal is able to act in relation to 'future' incentives. . . We summarise such a capacity by saying that the animal is able to construct internal models of its external environment and the occurrence of incentives therein.
>
> By postulating the involvement of such internal models of the world, we can see that motivation and goal-directed behaviour can attain some autonomy from stimulus control.

Toates (1986) makes much of the necessity to postulate internal representations of goals-to-be-achieved in accounting for purposive behaviour. He does not, however, distinguish between goal-achieving, goal-seeking and goal-directed systems. These can all give rise to apparently purposive behaviour. The problem is to determine what kind of system is operating in any particular case. I recognize that goal-directed systems may be capable of doing the job, but I will argue that there are alternative ways of doing the job, that goal-direction is unlikely in a complex motivational system, and that there are evolutionary reasons for supposing that humans have a predeliction for such teleological explanation.

It is sometimes argued that certain aspects of behaviour are in themselves, evidence of goal-directedness. These include, persistence, plasticity and disappointment. However, behaviour indicative of persistence and disappointment are easily accounted for without recourse to any notion of goal-directedness. When the consequences of behaviour differ from normal, the animal initially 'tries harder', persists, or shows 'behavioural inertia'. Such phenomena are commonplace, and various explanations have been offered. These include damming-up, positive feedback, and dishabituation (see Chapter 1 and 2 for some discussion of these topics). However, the notion of differing from normal may seem to require an explicit representation of the normal. This is a feature of some explanations, but it is not a necessary feature (see Mackintosh, 1983, pp. 225–8).

A classic feature of goal-directed behaviour is plasticity. This is the ability to choose among alternative routes to the goal according to the prevailing circumstances. Many have regarded plasticity as evidence of purpose (e.g. Braithwaite, 1946, 1953), and the subject has aroused a diversity of opinions. Many of the arguments concerning plasticity and teleology are reviewed by Woodfield (1976).

I am willing to accept plasticity as evidence that the behaviour has a goal, but because it can occur in both goal-achieving and goal-seeking systems, I do not see how it can be taken as evidence of goal-directedness. Plasticity shown by a goal-achieving system is nothing more that evidence of redundancy in the design of the system as a whole. Thus the embryo may take alternative developmental routes, a phenomenon called equifinality (Bateson, 1978). In this, and in other goal-achieving systems, natural selection has acted as a designing agent in providing the alternative routes.

Plasticity can also be shown by a goal-achieving automaton. For example, insects show plasticity of gait, and an individual can modify the pattern of leg movement according to the speed of locomotion, or in response to the loss of one or two legs. It has been discovered, however, that the co-ordination of leg movement follows a simple set of rules, which can account for a variety of gaits, even those shown by insects that have lost legs (Wilson, 1966). This is an example of a system designed (by natural selection) with built-in contingent behavioural alternatives.

In summary, I can see no compelling physiological or behavioural evidence or argument for the existence of the type of explicit representation necessary for true goal-directed behaviour. Moreover, there are alternative ways of accounting for apparently goal-directed behaviour, and it is to these that I now turn.

Functional Explanation of Apparently Purposive Animal Behaviour

In the study of animal behaviour two distinct types of explanation are employed. One is concerned with the way in which proximal causal mechanisms combine to control the behaviour of individual animals. The other accounts for behaviour in terms of hypotheses which purport to show how natural selection has, in the past, acted as a designing agent in shaping the evolution of behaviour. Such explanations are normally called 'functional' explanations.

In seeking explanations in terms of proximal causes, we are interested in aspects of animal behaviour that are far too complex to be accounted for in physiological terms. Although it is often possible to see how physiological processes are implicated in behaviour, it is doubtful that we will ever see a complete physiological explanation of the mechanisms controlling anything but the most simple aspects of animal behaviour. Even if the necessary feats of ingenuity and technology were possible, it is doubtful that we would want such a detailed account, any more than a programmer would want a detailed circuit diagram of his computer. Those in the business of repairing brains or computers, of course, need to have knowledge of the hardware, but those interested in behaviour find hardware-independent explanations are more useful.

In seeking to explain apparently goal-directed behaviour, we are not looking for a physiological account, but our explanation must be concordant with known physiology, and must be capable of refutation by physiological evidence. To attempt an explanation in terms of physiology would be to attempt at an inappropriate level of explanation, but at the same time any explanation we propose will inevitably have physiological implications which cannot be ignored (Noble, 1988).

The most common approach is to account for goal-achieving behaviour in terms of control systems theory, employing descriptive, analytical or computer simulation approaches (see McFarland, 1971; Toates, 1980, 1986, for examples of these approaches). This type of explanation is independent of the hardware employed by the animal. Essentially, it explains the behaviour by analogy. The problem with this approach is not in arriving at a model that can account for the behaviour, but rather in testing for the uniqueness of any model (McFarland and Houston, 1981). Inevitably there are other models that can do the job just as well. Rather than being satisfied with ad hoc simulation of purposive behaviour, it might be better to approach the problem from a design viewpoint.

Functional explanations are predicated on the assumption that natural selection acts as a designing agent in shaping the control mechanisms of physiology and behaviour. The design criteria are the features of the system upon which selection acts. Variation between individuals with respect to these features affects their fitness. Over long periods of time in a particular environment, the design is said to be optimized, because natural selection tends to maximize fitness, and fitness depends upon the design criteria (see Krebs and McCleery, 1984; Maynard-Smith, 1978; McFarland and Houston, 1981, for recent accounts of this approach).

The use of optimality principles in accounting for animal behaviour is analogous to the use of extremal principles in physics. A stone thrown through the air obeys a 'least action' law, minimizing a particular function of kinetic and potential energy. The stone behaves as if seeking an optimal trajectory, although its behaviour is in fact determined by forces acting in accordance with Newton's laws of motion. An animal may behave as if seeking an optimal trajectory through the state space (Sibly and McFarland, 1976; McFarland and Houston, 1981) but such an account says little about the mechanisms involved. Optimality principles are, in effect, functional explanations which describe how the animal ought to behave in order to attain some objective. For a causal explanation, the equivalent of physical forces must be established. These will presumably take the form of a set of rules-of-thumb governing behaviour, or of some equivalent mechanism.

Extremal principles may be regarded as teleological statements (Nagel, 1961), but the systems whose behaviour they describe would not normally be regarded as goal-directed. Many physical systems come to a natural equilibrium, at a minimal energy configuration, simply as a result of the resolution of counter-acting forces. To regard such minimal state as goals would be to trivialize the usefulness of the concept (see also Woodfield, 1976, pp. 67–68). A ball that comes to a rest after rolling down a hill would not normally be said to have reached its goal, or to roll down the hill in order to achieve a certain state. It is, nevertheless, goal-seeking in the sense described above. The behaviour of the ball conforms to an extremal principle which is analogous to the principles used to account, for example, for the profitability-maximizing behaviour of a foraging insect, fish or bird. In other words, there is an analogy between the extremal principles of physics and the optimality principles of biology (Rosen, 1967). This does not mean that the individual animal will always behave optimally in its natural environment. Genetic variation between individuals, the patchy nature of the environment, and evolutionary lag, all combine to make it very unlikely that the individual animal could ever be perfectly adapted to its niche. Nevertheless it is useful to imagine an animal that is perfectly adapted, because we can then specify what an animal would have to do to be perfectly adapted. The answer, in general terms, is that the animal would have to employ a form of dynamic optimization such that its behaviour always satisfied the optimality criteria embodied in a cost function (McFarland and Houston, 1981).

A cost function specifies the instantaneous level of risk incurred by (and reproductive benefit available to) an animal in a particular internal state, engaged in a particular activity, in a particular environment (McFarland, 1977). Thus the cost function is characteristic of the environment, in the sense that it reflects the selective pressures that moulded the optimality criteria during the course of evolution. The 'perfectly adapted animal', by deploying its behavioural options so as to continuously maximize a particular mathematical function achieves that behaviour sequence which maximizes its inclusive Darwinian fitness in the prevailing circumstances.

The individual animal may or may not be an optimizing machine in the sense that its behaviour conforms to a set of optimality criteria that are embodied within the animal itself. These criteria are referred to as the goal function of the animal (McFarland and Houston, 1981). Thus the goal function is envisaged as a property of the individual animal, and can be expected to differ from one member of the species to another. The individual animal is seen as an optimizing machine designed to maximize a particular mathematical function.

The goal function will include those properties of the animal which are relevant to the notional costs associated with the various causal states and with the various aspects of behaviour. These may include physical properties of the animal as well as evaluations represented in the brain. For example, a person travelling across town on foot has to make decisions as to when to rest, to walk, to jog, to pause etc. The goal function will include factors associated with the person's energy reserves, the risks of crossing roads etc. It would also include some more physical factors, such as the length of the person's legs (thus affecting the natural walking pace). Optimal decisions between jogging, walking, pausing and resting will inevitably be affected by all these factors, and will result in a trade-off between rates of energy expenditure and speed of travel.

We can now address the question of what it means to say that an animal is behaving optimally. There are three basic possibilities. Firstly, an animal may be supposed to be behaving optimally with respect to natural selection. This is tantamount to asking whether the behaviour conforms to a particular cost function. Secondly, an animal may or may not be an optimizing machine in the sense that its behaviour conforms with some goal function. A further distinction, which is discussed below, is that the animal may be behaving in an overall optimal manner or may only be behaving optimally with respect to a given set of constraints. Thirdly, an animal may appear to be behaving optimally if its observed behaviour can be shown to conform to a particular objective function. Once again, there may be a distinction between the overall behaviour and the constrained optimum. The question also arises as to whether an animal that can be shown to be behaving optimally with respect to an objective function, is necessarily behaving optimally with respect to a goal function.

An inevitable consequence of the optimality approach to animal behaviour is that there will be trade-offs among alternative courses of action. The logic of this conclusion is fairly simple. If the conditions relevant to only one activity pertain at a particular time, then the optimal policy is straightforward. The animal engages in that activity and optimizes its pattern of behaviour with respect to the use of energy, time, etc. For example, the optimal behaviour for a hungry pigeon faced with a source of food is to eat at a negatively accelerating rate (Sibly and McFarland, 1976; McCleery, 1977). If the conditions relevant to two (or more) activities apply simultaneously, then the animal has to choose between them. Moreover, its state when one activity is possible is not the same as its state when two activities are possible, and so the optimal behaviour for one activity in the presence of the other possibility is not the same as for that activity on its own.

An animal designed to make decisions entirely on the basis of trade-off principles will always be more efficient than an animal designed otherwise. The designer cannot avoid the issue, because the animal has to be able to change from one task to another. Psychologists and philosophers have avoided the issue in the past, because they have tended to concentrate on one task at a time. Their models were the equivalent of one-task robots. Optimality theory tells us that the best design is to continuously trade-off among all important criteria. The effect of this design when implemented in a robot will be easily recognizable. The robot will sometimes concentrate on one task until it is finished, but at other times it will interleave tasks. The recent applications of optimality thinking to animal behaviour indicate that that this is what animals do. Thus pigeons which are both hungry and thirsty, and placed in a position where they can work for food or for water, do not complete one task before starting the other. They interleave the two in a manner that can be accounted for in terms of optimization (Sibly and McFarland, 1976). Foraging sticklebacks change their pattern of predation when there is danger from predators. They achieve an optimal trade-off between attention to prey and predators (Milinski and Heller, 1978; Heller and Milinski, 1979). Similarly, great tits change their pattern of foraging when their territory is threatened by intruders. They increase their vigilance at the expense of their food intake rate (Ydenberg and Houston, 1986).

My claim is that the behaviour of individual animals (and people) is guided, not by any explicit goal-representation, but by myopic hill-climbing behaviour of the type described by Sibly and McFarland (1976). Each animal evaluates the situation for itself in accordance with its goal function (i.e. the goal function sets a value-scale on each relevant variable). This evaluation process forms the hill, which the animal then proceeds to climb, maximizing height subject to the prevailing constraints. Any change in the situation changes the nature of the hill (i.e. a re-evaluation takes place). By evaluation, I do not mean a cognitive process (necessarily), but merely a process that assigns new values to the variables and parameters involved.

Hill-climbing by gradient-maximizing is a means of achieving goals that does not involve any goal-representation. An argument sometimes levelled against the hill-climbing model is that it works only for short-term goals, because it is not capable of going down a small hill in order to climb a larger hill. This criticism is valid in a landscape in which there are many hills. In our landscape, however, *there is only one hill*. We are dealing with systems designed by natural selection, in which the overall objective is to maximize inclusive fitness. Natural selection is itself a goal-achieving process with a single objective. It is true that in the genetic landscape there may be more than one hill, and that individuals within a population may be subject to selective forces relevant to different hills. This argument is not relevant here, because we are interested in individuals. It is the individual's goal function that determines the nature of the hill. Goal functions are notional cost functions (see above), and it may be the case that there is bimodal variation among both the cost and goal functions relevant to the members of a population. This does not alter the fact that the goal function of an individual is immutable, and will have been designed as if there were a single hill in the genetic landscape.

An observer of animal behaviour may postulate that the behaviour of an animal obeys some extremal principle involving energy expenditure, foraging efficiency or Darwinian fitness. The observer may seek to find an objective function which embodies these optimality principles, and then wish to argue that the animal behaves optimally with respect to some internal goal function (McFarland and Houston, 1981). This does not mean, however, that the behaviour of the animal is necessarily goal-directed. It implies only that the animal is designed (by natural selection) in such a way that the (motivational) forces governing its behaviour result in behaviour that is optimal with respect to the goal function. The question of whether the animal's behaviour is goal-directed is a question that concerns the nature of the decision-making mechanisms.

Trade-off Processes

Animals may make decisions on the basis of rules-of-thumb, or on the basis of a cognitive evaluation of the situation. In a given case it may be hard to differentiate between these two types of mechanism. For example, we may observe a person crossing the road on the basis of a set of simple rules. Another person may cross on the basis of cognitive evaluation of the traffic situation. It would not be easy to discover which method was being employed in a given situation.

The question of whether or not the road-crossing behaviour is optimal is a separate issue. The rule-of-thumb mechanism may be well designed or badly designed. The cognitive evaluation may be equally good or bad in relation to the external circumstances (the cost function), or in relation to the person's internal criteria (the goal function). Even if it could be established which type of mechanism was being employed, and whether or not it was optimally employed, we still would not know whether or not a goal-directed mechanism was involved.

Part of the problem is that different evolutionary design strategies can arrive at behaviour patterns that are superficially similar. The strategy most frequently employed by human designers faced with a goal-achieving problem is to set up a representation of the goal-to-be-achieved and then devise various means of achieving the goal. The machine is able to monitor its progress towards the goal by comparing the goal-representation with feedback from the consequences of its own behaviour. This is a common design philosophy behind guided missiles, robots etc. It is a relatively easy design to implement, but this does not mean that it is the one employed in the evolution of goal-achieving behaviour in animals. As indicated above, there are various possible alternative design strategies, which do not involve explicit goal-representations. Moreover, as I argue below, the goal-directed model seems unlikely to work in a complex system.

In real life, the animal does not pursue its goal regardless of other environmental factors. All animal behaviour is a continual trade-off between competing and conflicting priorities. Let us now look more closely into the rationale behind the trade-off principle.

We can imagine a robot designed to carry out the cleaning and cooking chores in an American kitchen. The robot has to make decisions about its use of time; when to stop cleaning the floor, what chore to do next etc. A well-designed robot will embody a goal function (McFarland and Houston, 1981) which relates the different options to a common currency. That is, it specifies the extent to which adopting each option contributes to the maximized entity. We can suppose that the robot is designed to maximize some notion of efficiency, involving work done per unit time, per unit of energy expenditure etc. The goal function is a design feature of the robot, but this does not mean that it is explicitly represented in the robot's brain. Many aspects of the design are relevant to the goal function, such as the diameter of the wheels, length of arms etc. In other words the goal function is tacitly represented, and is an emergent property of the design as a whole.

The robot can behave absolutely optimally, with respect to its goal function, only if it has complete freedom of action. In reality there will generally be constraints upon the robot's behaviour. The amount of energy available may be limited so that energy spent on one activity is subtracted from that spent on another within a given time period. The rate at which chores can be performed may be limited both by the nature of the environment and by the physical features of the robot.

The state of the robot in conjunction with the goal function specifies the optimal course of action, but because of the constraints inherent in the situation, the robot may not be able to follow that course of action. Therefore we have to distinguish between the optimal solution to the problem presented by a particular situation, and the optimal attainable solution. The robot may be doing its best, but its best may not be the best possible behaviour relevant to the situation.

The robot is designed to operate in a completely standardized American kitchen. The kitchen is an unvarying environment, except in a trivial sense, so the robot does not have to adapt its behaviour by learning. Assuming the robot is optimally designed, its goal function will be identical to the cost function that is characteristic of the American kitchen. In this example, the cost function specifies the costs and benefits in terms of some measure of efficiency. The optimally designed robot is perfectly adapted to the American kitchen and will always carry out its duties in the most efficient manner.

Suppose, now, that the robot is transferred to a French kitchen and that the cost function characteristic of a French kitchen is different from that of an American kitchen. The robot will continue to behave in concordance with the notional costs and benefits embodied in its own goal function. It will still be an optimizing machine, but it will not be behaving optimally from the point of view of the French cost function. In French eyes it would not appear to be an efficient machine, and it would not do well in the French market.

When operating in the environment for which it is designed, the robot can maximize efficiency by sequencing its behaviour in an optimal manner, within the limits set by the constraints inherent in the situation. In an alien environment we can expect the robot to adapt to a small extent but its behaviour will fall short of perfect because of the constraints on its behaviour and because of the discrepancy between its goal function and the cost function characteristic of the situation.

These are circumstances that normally face an animal in its natural environment. There are always constraints upon its behaviour which necessitate a certain amount of budgeting of time and energy. The environment nearly always differs from that for which the animal evolved, because of the evolutionary lag (natural selection takes time to catch up with environmental changes) and competition (which displaces the animal from its preferred habitat). In addition, the evolutionary stable strategy may result in a sub-optimal solution to a particular problem (Dawkins, 1980).

Suppose we now ask how the robot is internally organized. The most obvious simple organization would be for the robot to have different programs for different chores. It would then have a set of criteria for deciding which program to implement at any particular time. The decision-making process would rely on information about the robot's internal state, including the state of its fuel reserves, its memory of tasks recently accomplished etc, and on information about its environment, including the untidiness of rooms, the time of day etc. The decision-rules would involve such factors as the urgency of different tasks, their fuel costs etc. The rules would also have to incorporate some optimizing principle. For example, suppose the robot rated each task on the basis of two independent indices: an index of task urgency and an index of fuel costs. We have to remember that fuel costs per unit time spent on a task will tend to be reduced as the total time spent is increased. This is because there will inevitably be a cost of changing between tasks. Thus, on the one hand it is better not to leave tasks unattended for too long (the urgency factor), while on the other hand, it is more fuel-efficient to change between tasks as little as possible (the fuel-cost factor). As in most decision-making problems there has to be some trade-off among opposed alternatives. The outcome of the trade-off depends upon the decision-rules, and the optimality criteria they embody. In Figure 5.3, for example, we see five tasks arranged on a graph of urgency index versus cost index. Which task should have priority? To answer this

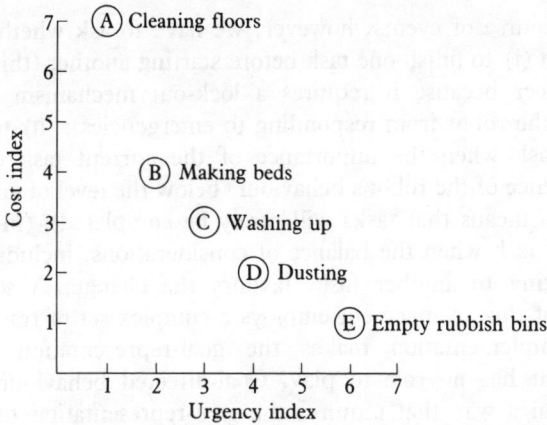

Fig. 5.3 Five household tasks arranged on a graph of urgency index versus cost index. The priority task is determined by a decision-rule. If the decision-rule maximized urgency index plus cost index, then the tasks would be ranked as follows: cleaning floors = 8 (7+1) > empty rubbish bins = 7 (1+6) > making beds = 6 (4+2), washing up = 6 (3+3), dusting = 6 (2+4). If the decision rule maximized urgency index times cost index, then the tasks would be ranked as follows: washing up = 9 (3×3) > making beds = 8 (4×2), dusting = 8 (2×4) > cleaning floors = 7 (1×7) > empty rubbish bins = 6 (1×6). Thus cleaning floors has priority on an additive decision-rule, while washing up has priority of a multiplicative rule. Other decision-rules are also possible.

question we would have to employ a specified decision-rule, as explained in the caption.

Having decided which task to engage in, the robot starts work. It is here that we might suppose the robot's behaviour to be goal-directed. According to this view, the robot would have an explicit internal representation of the goal-to-be-achieved and would monitor its progress by reference to this goal. Such a model might work satisfactorily if we consider only one task at a time. In real life, however, the robot is considering many tasks simultaneously. To handle this the model has to consider some form of competition and choice among alternative goals, but it is here that goal-directed theory starts to run into trouble (this is usually as far as the theory ever gets).

The problem becomes apparent when we consider what is to make the robot stop a particular task and start another. One scenario, with which the goal-directed approach has no difficulty, is that the robot changes to a new task if another goal suddenly becomes more important than the goal it is currently pursuing. When the baby spills ink on the carpet, the robot immediately recognizes this as an urgent task, drops whatever it is doing and deals with the emergency. In the normal course of events, however, we have to ask whether we want our robot (i) to finish one task before starting another (this is difficult to engineer because it requires a lock-out mechanism which may prevent the robot from responding to emergencies); (ii) to change to a new task when the importance of the current task drops (as a consequence of the robot's behaviour) below the level of an alternative goal (this means that tasks will rarely be completed); (iii) to change to a new task when the balance of considerations, including the cost of changing to another task, favours the change. A sophisticated version of this arrangement employs a complex set of trade-off rules. Their implementation makes the goal-representation redundant, because it has no role to play. Goal-directed behaviour is defined (above) in a way that requires the goal-representation to be instrumental in guiding the behaviour. If the behaviour is determined entirely by trade-off considerations, then the goal-representation has no role to play, it is not guiding the behaviour, therefore the behaviour is not goal-directed.

The animal, or robot, surveys the options available at each point in time. These change according to the external circumstances and the animal's internal state. The options present themselves already evaluated in various ways. In the case of our robot, the evaluations will be preprogrammed. In the case of an animal, they will be partly preprogrammed and partly learned. What is evaluated is not the precise situation, but the operating range of the relevant variables. For example, an important variable in a robot is the state of the fuel reserves. In weighing the options the robot will take account of this value, as well as the value attached to the variable indicating the proximity of the fuel source. Thus the priority given to the refueling option in competition with other behavioural options will depend jointly on the state of the reserves and the availability of fuel. Exactly how these two factors combine (i.e. the combination rule of McFarland and Houston, 1981) is a matter of design.

Even if there were an explicit representation of the robot's goal-to-be-achieved at any particular point in time, this representation could not have an active controlling role in the sense required for goal-directed behaviour. An essential feature of an active control system is that it corrects for the disturbing effects of extraneous influences of the output, as illustrated in Figure 5.4. This is usually achieved by careful design of the dynamics of the controlling system (see McFarland, 1971, chapter 5, for a more technical discussion). If the design is successful then all but transient influence of extraneous inputs is eliminated, and there is no trade-off with these variables. If

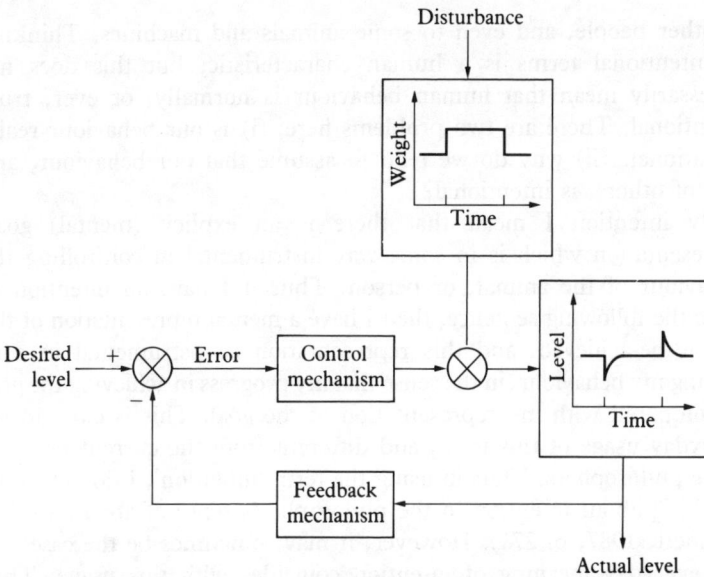

Fig. 5.4 An example of the way in which a goal-directed system is designed to counteract disturbance. Imagine a waiter carrying a tray of mugs of tea. His goal is to maintain the tray at a constant level. At a particular point in time a mug is added to the tray (see disturbance) increasing its weight. This causes a sudden dip in level which is quickly corrected. At a later time a mug is removed, reducing the weight. This causes a sudden rise in level, which is quickly corrected.

trade-off is desirable, then the degree of control must be relaxed, because the two design criteria are incompatible. Once trade-off is permitted, then the role of the goal-representation is no different from the role of any other input variable, and the system has ceased to be a goal-directed system.

This part of the argument may be summarized as follows: trade-off among possible activities is an optimal design feature of any complex behavioural system, be it robot or animal. Such trade-off is incompatible with the essential design features of goal-directed behaviour, because active (goal-directed) control systems are designed to eliminate the influence of extraneous variables, whereas the essential feature of trade-off is to allow such influences.

Intentional Behaviour

We normally think of intentional behaviour as typically human, premeditated, cognitive behaviour. We attribute intentional behaviour to other people, and even to some animals and machines. Thinking in intentional terms is a human characteristic, but this does not necessarily mean that human behaviour is normally, or ever, truly intentional. There are two problems here: (i) is our behaviour really intentional? (ii) why do we tend to assume that our behaviour, and that of others, is intentional?

By intention I mean that there is an explicit (mental) goal-representation which is in some way instrumental in controlling the behaviour of the animal, or person. Thus if I have an intention to write the following sentence, then I have a mental representation of the goal-to-be-achieved, and this representation is instrumental in controlling my behaviour, in the sense that my progress in achieving the goal is compared with my representation of the goal. This is close to the everyday usage of this term, and different from the current usage of some philosophers. Thus in using the term 'intention', I do not imply anything about intention in the philosopher's sense of aboutness (see Dennett, 1987, p. 271). However, it may sometimes be the case that the everyday meaning of intention coincides with this usage. Thus when Dennett (1983) refers to the monkey: 'Tom wants to cause Sam to run into the trees. . . On this reading the leopard cry belongs to the same general category with coming up behind someone and saying "Boo!"', he is implying that Tom has the intention of scaring Sam into the trees in the sense that this goal is represented in Tom's mind before he gives the alarm. Like the goal-representations used in modelling animal behaviour, the intention-representation acts as a standard of comparison for feedback from the consequences of behaviour. Thus Tom, seeing that Sam has not fled into the trees, gives the alarm call. When Sam has fled into the trees, Tom does not give the alarm call (McFarland, 1983).

Thus an intentional system is a form of goal-directed system in which there is a mental representation of the goal-to-be-achieved. Despite the popularity of this model, there are problems about its veracity. The first problem is one of evidence. Apart from introspection, we do not have, and probably never will have (Noble, 1988) scientific evidence for the existence of the type of explicit (mental) representation necessary to produce intentional behaviour. Indeed, as I argue earlier in the chapter, we do not have behavioural or physiological evidence for the type of explicit goal-representation necessary for simple goal-directed behaviour. Moreover, if we believe that our behaviour is the result of complex trade-offs among various possible courses of action, then there is no room for intentional behaviour. In other words, as in the case of simple goal-directed behaviour outlined

124

above, the trade-off model is incompatible with the intention-model. However, in addition to its supposed role in guiding behaviour, this representation may have other roles, and is probably particularly important as a basis for language, as I argue in Chapter 6.

The second problem is that when we see goal-achieving behaviour it is natural for us to assume that it is goal-directed. I suspect that we make this type of assumption because we normally think and communicate in teleological terms. Introspection tells us that much of our own goal-achieving behaviour is intentional, and we tend to assume that the behaviour of other people, of some animals, and even of some machines, is similar. Our introspection may be misleading, for good evolutionary reasons which I discuss in Chapter 6. Even if our introspection is a good guide to our thinking on some teleological matters, it is certainly a poor guide on others. More than a hundred years after the publication of Darwin's *Origin of Species*, it is still necessary to convince many people that evolution is not purposive, but takes place as a result of a simple goal-achieving process, called natural selection. So pervasive is the tendency to assume that goal-achieving phenomena are purposive, that some biologists (e.g. Dawkins, 1986) go to considerable lengths to persuade the layman otherwise. It is not self-evident to me that introspection can be relied upon to point the way to the workings of human, or animal, motivation. I find it disturbing that so many writers on teleological matters take as their starting point the 'evident' goal-directedness of human behaviour. Of course, they may not all mean that the goals are explicitly represented in the brain, as required by my definition of goal-directedness. In some cases the term is used loosely, without much thought as to how the goal is to be achieved.

The issue, as I see it, is as follows: We have no hard evidence for representations of goals-to-be-achieved. There are some persuasive arguments to the effect that such representations must exist, at least in language using species (e.g. Fodor, 1987; Dennett, 1987), but to argue that explicit representation must have a role in linguistic behaviour is not the same as arguing that ordinary behaviour is goal-directed. I suggest that, contrary to the tenets of action theory, explicit representations (if they exist) do not control our behaviour, or that of animals, although they may have a role in communication (see Chapter 6). Indeed we tend to assume that our own, and other people's, behaviour is intentional because of the role intention plays in our linguistic behaviour, and therefore in our thinking.

CHAPTER SIX

Cognition in animals

We are animals that are capable of cognition. When we think about cognition in other animals, we tend to think of abilities that are similar to our own cognitive abilities. We recognize that such anthropomorphism is not good scientific practice, when we are dealing with animal perception, motivation or learning, but when it comes to animal cognition we seem prepared to abandon this dispassionate view.

One argument often advanced in support of animal cognition is that some animals are capable of sophisticated and well adapted behaviour. 'The flexibility and appropriateness of such behaviour suggest not only that complex processes occur within animal brains, but that these events may have much in common with our own mental experiences' (Griffin, 1976, p. 3). I fail to see the logic of this argument unless it is based on some notion of evolutionary continuity. Even here, it seems to me, we are treading on dangerous ground. Evolution is a double-edged sword. On the one hand, we can argue that man is an animal, and should not assume that he is all that different from other animals. 'Must we reject evolutionary continuity in order to preserve our gut feeling of human superiority . . .?' (Griffin, 1981, p. 112). On the other hand, an evolutionary argument would lead us to suppose that each species has to solve the problems appropriate to its niche. Just as the sensory world of bats is very different from our own, so may be their mental experiences and cognitive abilities (Nagel, 1974). Thus, as a justification for supposing that animals have mental experiences similar to our own, the evolutionary argument cuts both ways.

Another approach is to take a phenomenon, such as deception, which is typically thought of as a cognitive aspect of human behaviour, and look for its counterpart in other species. A well known example is the story of the fox that cried wolf.

> On the second day, a cub came with the adult to our tent site, and still another day later a second cub came with the adult to the camp. On this day we were able to make the following

observations: the young foxes would not take food out of the hand, and would not approach closer than 1 to 1.5 m. Therefore, the older fox got the most and the biggest morsels. When the adult left with a large chunk of cheese, a young fox noticed it and ran after her and jumped barking and biting at her, whereupon she dropped the morsel. The young fox turned to the cheese morsel, rotated its rump and urinated towards the adult. Then it began to eat the morsel. The adult took a few steps to the side and watched. Suddenly she whirled her snout upwards and repeatedly gave high pitched warning calls. The young fox immediately dropped the food morsel and disappeared into the rocks. The vixen then grabbed the fallen food morsel and ate it . . . In the following days, the adult fox again drove the youngsters away several times to obtain the foodstuff by using the warning call, without a noticeable danger being present (von Ruppell, 1986, pp. 177–8).

Some would call this an anecdote, others evidence of deception practised by animals. A recent book contains accounts of many such observations, on animals ranging from fireflies to chimpanzees and humans (Mitchell and Thompson, 1986). The observations themselves are often interesting, but their treatment presents problems. Like language, deception can be defined in such a way that it can be seen in almost every species studied, or it can be confined to humans. This seems to be a matter of personal preference, and not an issue worth arguing about. The interesting issue is what criteria can be used in identifying deception as truly cognitive. We may be ready to accept that the bluff and deception displayed by stomatopods (mantis shrimps) in their aggressive encounters does not involve cognition, simply because they are invertebrates (Caldwell, 1986). For mammals the question is not so clear cut. Did the vixen figure out how to trick her cubs into dropping the food, or had she simply learned a circus-like trick to obtain food?

A third approach is to ask what animals might gain from cognitive abilities. This is essentially a functional approach, and not one likely to throw much light on cognitive mechanisms. Nevertheless, it is often helpful in investigating mechanisms to have some understanding of their evolution and functional role.

In this essay I will explore the possibility that humans are designed (by natural selection) to assume that deceitful (and other) acts by others are intentional. I will do this by taking a functional approach to cognition.

Functional Aspects of Learning

If the environment never changed, animals would gain no benefit by learning. A set of simple rules designed to produce appropriate behaviour could be established by natural selection and incorporated, as a permanent feature of the animal's behaviour-control mechanisms. There are some features of the environment that do not change, and we usually find that animals respond to these in a stereotyped manner. For example, gravity is a universal feature of the environment, and antigravity reflexes tend to be stereotyped and consistent.

Some features of the environment change on a cyclic basis, engendering circadian, circalunar, or circannual rhythms of physiology and behaviour. There is no real necessity for animals to learn to adapt to such environmental changes, because the necessary responses can be preprogrammed on the basis of an endogenous biological clock (McFarland, 1985). In general, environmental changes that are predictable over long periods of time can be handled by preprogrammed changes in the animal's makeup. We are all familiar with seasonal changes in motivation, based on hormonal and acclimatory changes, with endogenous tidal rhythms of activity in marine and littoral animals, and with endogenous rhythms of sleep and wakefulness (see Chapter 1).

Similar principles apply to changes that occur throughout the animal's life history, since many of these are also evolutionarily predictable. In some, the behaviour changes by maturation, without learning being involved. In other cases, the behaviour may change as a result of preprogrammed learning that occurs at a particular stage of life, more or less independently of the individual's circumstances. Thus human children have an immense facility for language learning between the ages of 2 and 7, picking up whatever language they hear around them. Similar phenomena occur in bird-song learning. Through imprinting, many species learn the characteristics of their habitat, their parents or their siblings. Such learning is essentially preprogrammed, occurring during a sensitive period of development, and resulting in an attachment to whatever salient stimuli were provided by the prevailing environmental circumstances (McFarland, 1985). Whenever the problems facing the juvenile animal differ slightly from generation to generation, there will be an advantage for individuals that are predisposed to learn certain types of things, such as the nature of the habitat, close relatives or food sources. This type of learning is a form of contingent maturation, the course of development being contingent upon certain experiences.

128

Unpredictable environmental changes that occur within an individual's lifetime cannot be anticipated by preprogrammed forms of learning or maturation. The individual must rely upon its own ability and experience in adapting to such changes. There are various simple forms of learning, such as habituation, and stimulus-substitution, which would seem to provide a limited means of adjustment, but the ability to modify behaviour appropriately in response to unexpected environmental change calls for some form of intelligence.

It used to be thought that learning by association (classical conditioning), or through the consequences of behaviour (instrumental learning) were fairly automatic processes, requiring no cognition. This view is now disputed, and it is claimed by some (e.g. Dickinson, 1980) that some form of cognition is required for associative learning. Be that as it may (for the time being), we should be clear that, in talking about cognition, we are talking about a form of phenotypic adaptation. This may seem obvious, but we have to be careful here, because the apparatus responsible for cognition is a product of genotypic adaptation. In taking an evolutionary perspective we need to define cognition in a way that is both free from anthropomorphic overtones and distinct from other forms of phenotypic adaptation. It is here that problems arise.

Intelligence and Cognition

Romanes (1882) defined intelligence as the capacity to adjust behaviour in accordance with changing conditions. His uncritical assessment of the abilities of animals provoked a revolt, led by Lloyd Morgan (1894) and developed by the behaviourists, who attempted to pin down animal intelligence in terms of specific abilities. Modern thinking is more in line with Romanes view.

At the end of his historical analysis of the changing views on the nature of intelligence, Tuddenham (1963) comes to the conclusion that 'intelligence is not an entity, nor even a dimension in a person, but rather an evaluation of a behaviour sequence (or the average of many such), from the point of view of its adaptive adequacy. What constitutes intelligence depends upon what the situation demands. . .' (p. 517). This view is endorsed by Hodos (1982) who promotes the idea of animal intelligence as an abstract characterization of the individual's behavioural responses to pressures from the environment. He points out that animal intelligence should not be judged by comparison with human intelligence, because human intelligence is

129

special as a result of 'language and related cognitive skills, which permit us to communicate not only with each other, but with past and future generations' (p. 37). Moreover, animals have special capabilities which suit them to their way of life.

An extreme version of this view can be found in this discussion of intelligence in invertebrates.

> The imposition of vertebrate biases on invertebrates predisposes thinking that what is intelligent behaviour for the vertebrate must be a useful and desirable capacity for the invertebrate. An example, the capacity to associate stimuli is probably one of the basic requirements for vertebrate intelligence but what is intelligent for the vertebrate need not represent intelligence for the invertebrate; in annelids the existence of such learning may prove to be of much more significance to the animal behaviourist than to the worm. While it may well be that the worm CAN learn associatively the demands of their normal environments seldom, if ever, actually require that they do so. (Corning, Dyal and Lahue, 1976, pp. 216–17).

This approach to animal intelligence is similar to the functional analysis practised by ethologists and behavioural ecologists (Houston and McNamara, 1988; McFarland and Houston, 1981; Stephens and Krebs, 1986). Functional analysis attempts to specify the costs and benefits that accrue to different aspects of behaviour, and attempts to formulate the best behavioural strategy and tactics that will enable the animal to adapt to the current circumstances. It does not in any way specify how the optimal strategy is to be achieved. There are many clever things that animals can do, that we do not normally attribute to intelligence. The feats of visual processing carried out by the hawk's eye and brain, the navigational abilities of pigeons, and the fantastic co-ordination of monkeys in their arboreal locomotion, are products of innately pre-wired subsystems which we do not usually regard as intelligence (Rozin, 1976a). Part of the problem is that we tend to confuse intelligence and cognition. The system controlling spatial orientation in humans is a highly sophisticated, pre-wired adaptive control system (Howard, 1982), which would seem extremely intelligent if performed by a robot. The control of spatial orientation requires no cognition (as far as we know), but the behavioural outcome does appear to be intelligent. We must distinguish between cognition, a possible means to an end, and intelligence, an assessment of performance judged by some functional criteria.

Thus cognitive ability is not merely the ability to produce clever solutions to problems posed by the environment. Nor is it an ability that necessarily increases along an evolutionary scale. Evolution does not progress in a linear manner. It diverges along a multiplicity of routes, each involving adaptation to a different set of circumstances. This means that animals may exhibit considerable complexity in some respects but not others and that different species may reach equivalent degrees of complexity along different evolutionary routes. For example, proficiency in the visual recognition of objects, regardless of their relative spatial orientation is a feature of several tests of intelligence and aptitude in humans (Petrusic, Varro and Jamieson, 1978). When the same tests are given to pigeons, it appears that pigeons are able to solve this type of problem more efficiently than humans, though apparently by a different method (Hollard and Delius, 1983). In humans performance in the tests is taken as an index of cognitive ability, but in pigeons it probably should not be so taken.

The fact that some species do have sophisticated hard-wired mechanisms for dealing intelligently with certain well specified situations suggests that the alternative, cognitive, approach is either more expensive, or less efficient in some way. It might be that a cognitive mechanism required more brain than a hard-wired solution, or it might be that the cognitive approach is more fallible. To be able to tackle this kind of problem we need to have some idea of the nature and limitations of cognition.

The Nature of Cognition

Cognition involves learning and thinking processes which are not directly observable, but for which there often seems to be indirect evidence. Evidence for what precisely? In asking this question we discover that cognition is notoriously difficult to define. For Toates (1986) it seems sufficient that 'information may be stored in a form not directly tied to behaviour. Such information may be exploited in different ways according to context. . .' (p 14). This seems to me to be more like a definition of memory, but Toates goes on to endorse the definition of Menzel and Wyers (1981):

> Our principal definition boils down to a negative one: cognitive
> mechanisms may be assumed to be necessary when all other
> alternatives may reasonably be rejected – that is, when
> description in terms of Newtonian or sequential stimulus-
> response, input-output relations falls short and the animal seems
> to be filling in certain information gaps for itself or going beyond

the information or stimuli that are immediately available in the external environment.

There are a number of problems here. Unless we are to take the stimulus-response notion literally, I suspect that it will always be possible for the ingenious computer programmer to come up with an 'input-output' model that does the job. Like Dennett (1983), I see the behaviourist and cognitive explanations as possible, and plausible, alternatives. The question is partly one of which is correct in a particular case (a problem solved cognitively by one animal might be solved non-cognitively by another), and partly one of which type of explanation is to be preferred.

For Honig (1978), cognitive terms can be used to describe sets of behavioural observations, and if these terms possess attributes beyond those specified by the defining observations, they can be theoretical terms as well. In this way cognitive terms acquire explanatory capacity. This is true of much behavioural terminology, such as motivation, expectancy etc. It does not tell us what is distinctive about a cognitive process.

It is not clear to me why going beyond the information that is immediately available should imply cognition. In assessing stimuli, animals often do this. Their perceptual systems are built to make certain assumptions. For example, navigating bees can sense the direction of the sun even when it is obscured by clouds, but the information they obtain can be ambiguous, there being two patches that look the same, positioned symmetrically with respect to the sun. The bees overcome this problem by acting as if the patch they see is the one to the right of the sun (Gould, 1980). Thus the bees go beyond the information immediately available by adopting a convention. This need not imply cognition, since it could be the result of a simple hard-wired rule. Moreover, animals do not react solely on the basis of external stimuli, but are also influenced by internal factors, such as their endogenous clock, memory of past situations etc. Thus they inevitably 'interpret' external stimuli in terms of their internal state, but again, there is no compelling reason to regard this as a form of cognition.

What many attempts to define cognition seem to hint at is that we should regard as evidence of cognition anything remotely akin to human cognition. This is unacceptable, if only because our diagnosis of human cognition may be incorrect (as I suggest below). It is merely a covert way of adopting an anthropomorphic posture, a posture that we reject when investigating other aspects of behaviour.

In inquiring into the functional role of cognitive processes, we are looking for roles that are not covered by the conventional mechanisms of learning and adaptation. There are a number of possibilities here. Firstly, if cognition provides a way of coping with unpredictable aspects of the environment (see above), we might expect that the individual's perceptual evaluation of the situation be flexible. The stimulus situation would not be automatically categorized into sign stimuli, or associated with a set of conditioned reflexes. It must be evaluated in some way that poses a problem for the animal, otherwise there is no call for a cognitive solution. Secondly, the animal's response to the situation must be novel, in the sense that the individual has not hitherto typically given the same response to the same situation. Thirdly, the animal's evaluation of the consequences of its own behaviour must somehow deal with the fact that an unusual situation has been responded to, or a novel response made. The problems posed by unusualness and novelty do not arise in the cases of preprogrammed learning, maturation, or physiological adaptation, but if the animal is to adapt in an unpredictable environment, then its responses must in some sense be constructive.

An animal that interferes with its environment, without knowing the consequences is conducting an experiment, except in those cases of preprogrammed behaviour, where the consequences are evolutionarily predictable, and the individual does not have to monitor them. The idea that some animals may form hypotheses about the consequences of their own behaviour has long been a theme in the psychological literature. I am not concerned here with the evidence for such a view, but rather with the scope and limitations of such a process. To what extent is cognition limited by the animal's ability to handle information?

If the cognitive animal is like an experimenter, detecting new features of the environment, responding in a tentative manner, and drawing conclusions on the basis of the consequences, then we might expect it to be subject to some of the limitations placed upon human experimenters. Two important concepts here are observability and controllability. These concepts apply in two types of situation. One is the situation in which an investigator is attempting to understand the workings of a system. Certain aspects of the system may not be observable, and certain aspects may not be controllable by the investigator. This unobservability and uncontrollability pose problems in the interpretation of scientific data. The other situation, which I will discuss later, concerns the relationship between the animal (or human) and its own behaviour.

133

The concepts of observability and controllability were first proposed by a systems analyst, called Richard Kalman, in 1963. They have proved to be very important in control systems engineering. In 1968 Kalman suggested that these concepts might be important in biology. A system is said to be completely controllable if it is possible to move any state to any other state in a finite time by a suitable choice of input. A system is said to be completely observable if observation of the inputs and outputs for a finite time enables the state to be determined. Kalman established that the input-output relations determine only that part of the system that is completely controllable and completely observable.

McFarland and Sibly (1972) discuss the problem of observability in their treatment of the question of how the nature of the environment may constrain behaviour. The animal may have a goal which the environment does not enable it to obtain. In making a 'best' attempt to reach the goal it may behave in exactly the same way as an animal which can reach its goal. Thus, observable motivational commands (instructions to achieve a certain goal-state) are in one-to-one correspondence with their optimal consequences, whereas the relationship between unobservable commands and their consequences is equivocal. In other words, from the experimenter's viewpoint, the state of the animal can sometimes be deduced from behaviour, but sometimes, under apparently the same conditions, it is not observable. Similar conclusions have been drawn from simulation of newt courtship (Houston, Halliday and McFarland, 1977), which seems to be designed to obscure the true motivational state. Indeed, as we see below, this is a common design feature of animal communication.

The experimenter, in trying to explain, or build a model of, behaviour, faces a serious difficulty when the system is not completely observable. A model that contains no redundant elements is said to be minimal. A model that is the only one that can account for the data is said to be unique. Kalman (1963) showed that a model is minimal if and only if it is completely controllable and completely observable. Moreover, a minimal model is unique. Kalman's results apply only to linear systems, but they can be extended to non-linear systems by making use of the close correspondence between the dynamical systems theory, which Kalman used, and the theory of finite automata. Kalman (1963) compared his theorems with Moore's (1956) results on automata, and Arbib (1966, 1969) has investigated this relationship in considerable detail. The advantage of finite automata theory is that the question of linearity does not arise unless extra structure is added, that is, unless the system switches between modes which have different structures (see McFarland and Houston, 1981).

I do not propose to enter into details here. I hope that it will be sufficient for our purposes to note that the concepts of observability and controllability apply to those non-linear systems that can be analysed in terms of finite automata theory. From the experimenter's viewpoint, systems that are partially unobservable and uncontrollable are difficult to investigate. Any model made of such a system is inherently unreliable. However good it seems at predicting the behaviour of the system, there will always be another model that is equally good, because the model of a partially uncontrollable and observable system can never be unique.

The Animal as an Experimenter

Just as the human experimenter investigates a system and draws certain conclusions, so the cognitive animal attempts to gain knowledge by investigating its environment. It will often be the case that the consequences of the animal's investigative behaviour are not fully observable by the agent responsible for the behaviour. This inability to fully monitor the consequences of the animal's own behaviour may mean that some aspects of the behaviour are not completely under the animal's control. In a sense, an animal attempting to obtain feedback from its environment is in a similar position to that of an experimental scientist. Both try to obtain knowledge about the environmental situation by interpreting the consequences of their own behaviour. Just as the scientist cogitates about the information gained from experiments, so the cognitive animal must rely on information gained from experience.

Let us take a simple example from human behaviour. In learning a skill, such as that involved in golf, the person may have a motor routine appropriate to the play of a particular stroke. But the player also needs a subroutine which executes the necessary skilled motor movements. Having played the stroke, the person may have very little information as to what exactly went wrong with the stroke, although it is usually possible to judge how good, or bad, the shot was. One of the factors that makes this kind of skill difficult to learn is the poverty of feedback about the consequences of the behaviour. The information about the exact co-ordination and movement of the muscles is not readily available to the brain, and the player does not really know what actually happened. That the player has little information about the consequences of the behaviour, is another way of saying that the behaviour is relatively unobservable. An extreme case would be scalar information, such as 50 yards short, 20 yards short etc. The player now has little information by which to alter the skill-subroutine. In other words, the execution of the stroke is relatively uncontrollable. There may be an indication as to how much the stroke should be modified, but little indication as to how it should be modified. It is not very surprising, therefore, that golf is a difficult game to learn.

Knowledge of how to behave in various situations can be stored in a number of ways:

(a)   It can be inherent in the motor mechanisms themselves. Thus a decapitated hen knows how to run by virtue of the way in which the peripheral mechanisms are put together.

(b)   It can be represented in the brain in a form that can be read out as a set of instructions designed to execute a particular behaviour sequence. Many skilled movements, such as handwriting, are controlled directly by preprogrammed signals generated in the brain, no feedback being involved. The old Chinese bamboo painters trained themselves to use the brush in one movement, because retouching was impossible in their material (McFarland, 1971).

(c)   It can be represented in a manner such that it can be made available for use in many types of behaviour. Thus honey bees can make use of information provided by the dance of foragers on returning to the hive. This information can be used by bees leaving the hive to locate sources of food, of water, of propolis (a type of sap used to caulk the hive), and of possible new hive sites at swarming time (Gould, 1981). Information about the direction and distance must be represented, at least temporarily, in the outgoing bees, which navigate on the basis of the information conveyed by the scouts.

The question of how animals organize their knowledge of the consequences of behaviour has long been the subject of controversy. Dickinson (1985) has highlighted the two dominant opinions:

(a)   The mechanistic stimulus-response view, according to which a particular behaviour is either an innate or an acquired habit which is simply triggered by the appropriate stimulus.

(b)   The 'teleological' view, which maintains that some activities purposive actions controlled through the current values of their goals through knowledge about the instrumental relations between the actions and their consequences.

A number of problems are raised by the fact that Dickinson (1985) portrays rather extreme versions of these two different views. He contrasts 'responses' and 'actions'. The former is elicited by stimuli, and (by implication) the animal gains no feedback about the consequences of the response. The latter is 'controlled at the time of performance by the animal's knowledge about the consequences of this activity. In other words, the teleological model claims that at least 'some behaviour is truly purposeful and goal-directed. .' (p. 67). One issue here is whether behaviour, as 'controlled at the time of performance' is goal-directed or not. As we saw in Chapter 5, I have argued that it is not.

Another issue concerns 'goal revaluation'. Dickinson (1985) maintains that the type of control (stimulus-response or goal-directed) over any particular activity can be determined by a goal revaluation procedure. Specifically, if an animal's behaviour changes appropriately following an 'alteration of the value of the goal or reward' without further experience of the instrumental relationship, the behaviour should be regarded as a purposive action. I do not dispute, in the face of Dickinson's persuasive evidence, that some form of revaluation occurs. The question is whether 'goals' are revalued or whether some other entities are revalued. There is also a problem as to the extent to which revaluation should be considered as evidence for cognition.

These issues are best illustrated by discussion of one of Dickinson's own experiments. A crucial experiment was carried out by Adams and Dickinson (1981a), who trained hungry rats to lever press for one of two types of food (sugar or chow) while they received the other on a schedule designed to ensure that lever pressing and food presentations were uncorrelated. Thus for sugar the rewards were contingent on lever pressing, while for chow they were non-contingent. In another group of rats the roles of sugar and chow were reversed. They then devalued the contingent foods (counterbalanced design) for one group of rats (A), while maintaining the value of the non-contingent foods. For another group (B), the non-contingent foods were devalued and the value of the contingent foods was maintained. Food was devalued by allowing the rats to have access to the food in the absence of the lever, and injecting them with lithium chloride soon after they ate the food. The lithium chloride makes the rats mildly ill and induces an aversion to the food with which it is associated.

Both (A and B) groups were then given an extinction test in the lever-pressing situation. The result was that the contingent (A) group pressed at a lower rate in the extinction test than the non-contingent (B) group. To check whether the aversion procedure had been differentially successful, reacquisition tests were then given. These showed that both the contingent foods in group A and the non-contingent foods in group B had lost their capacity to act as rewards.

Dickinson's (1985) interpretation of these results is that the animals for which the contingent food is devalued (group A) perform less vigorously in the extinction test because the goal (obtaining this food) had been devalued. The animals for which the non-contingent food is devalued (group B) perform more vigorously because their goal (the contingent food) had not been devalued. The reacquisition tests show that the food itself is devalued in both groups. His conclusion is that 'performance of this particular instrumental behaviour really does seem to be controlled by the knowledge about the relation between the action and the goal' (p. 72).

An alternative possibility is that the consequences of the behaviour of pressing the lever, rather than the food or the goal, is revalued. In the one case (group A) pressing the lever is associated with aversive food, while in the other case (group B) it is not associated with aversive food, because the non-contingent food was paired with lithium chloride in this group. Thus for group A the value of the consequences of lever pressing behaviour are reduced in value relative to other possible activities (hence the lowered rate of pressing during extinction). For group B the consequences of lever pressing are not reduced in value, and so the rate of pressing during extinction is higher than in group A. This view would be consistent with the hill-climbing model outlined in Chapter 5, in which goal-directedness plays no role. For this reason I do not agree that the Adams and Dickinson (1981a) experiment necessarily provides evidence for goal-directed behaviour. In other words, it remains possible that lever pressing for food is not an intentional action.

There are two different issues here. There is the question of whether the animal's behaviour is purposive (or goal-directed as defined in Chapter 5), and there is the question of whether the animal's behaviour involves cognition. For Dickinson (1985) the answer is clearcut.

From my point of view . . . the main implication of our results concerns the nature of the cognitive processes controlling instrumental behaviour. My colleagues and I [Adams and

Dickinson 1981b; Dickinson, 1980; Mackintosh and Dickinson 1979] have argued that teleological control of instrumental behaviour cannot be explained, at least at the psychological level, in terms of internal associations which have just excitatory or inhibitory properties. Rather, we argue that the knowledge about the goal-action relation must be encoded in a propositional-like form so that it can be operated on by a practical inference process to generate the instrumental performance (p. 78).

The problem here is to know whether the cognition issue is necessarily linked to the goal-direction issue. In order to address this problem without complications arising from considerations of particular species, I propose to take up the robot analogy outlined in Chapter 5. We imagine a robot designed to carry out the cleaning and cooking chores in an American kitchen. The robot has to make decisions about its use of time; when to stop cleaning the floor, what chore to do next etc. A well-designed robot will maximize some notion of efficiency, involving work done per unit time, per unit of energy expenditure etc.

When operating in the environment for which it is designed, the robot can maximize efficiency by sequencing its behaviour in an optimal manner, within the limits set by the constraints inherent in the situation. In an alien environment we can expect the robot to adapt to a small extent but its behaviour will fall short of perfect because of the constraints on its behaviour and because of the discrepancy between its goal function and the cost function characteristic of the situation.

In considering how the robot is internally organized, I rejected the idea that its behaviour is goal-directed, in the sense that there is an internal representation of a goal-to-be-achieved that is instrumental in controlling the behaviour of the robot. My arguments for this are outlined in Chapter 5.

The robot is designed to carry out a number of tasks and to schedule them in an efficient manner. This multi-task requirement necessitates a complex set of trade-off rules. Their implementation makes the goal-representation redundant, because it has no role to play that is different from the role of any other input to the behaviour control system. Goal-directed behaviour is defined in a way that requires the goal-representation to be instrumental in guiding the behaviour. If the behaviour is determined largely by trade-off considerations, then the goal-representation has no controlling role to play. It is not guiding the behaviour, therefore the behaviour is not goal-directed. However this does not mean that there can be no explicit representations. In thinking or talking robots, the goal-representation may have a role to play, though not a behaviour-guiding role (see below).

My claim (in Chapter 5) is that the behaviour of individual animals is guided, not by any goal-representation, but by myopic hill-climbing behaviour. Hill-climbing by gradient-maximizing is a means of achieving goals that does not involve any goal-representation. The animal, or robot, surveys the options available at each point in time. These change according to the external circumstances and the animal's internal state. The options present themselves already evaluated in various ways. So far, in the case of our robot, the evaluations are entirely preprogrammed. In the case of an animal, they will be partly preprogrammed and partly learned, so we should now consider what would be involved if the robot were to update its evaluations by learning.

In adapting to the new environmental circumstances, the robot could learn to alter the constraints on its behaviour, so as to perform better in terms of its own goal function. It could not learn to alter its own goal function to make it more like the cost function of its new environment. Although an animal which could do this would greatly increase its overall fitness, we have to remember that all learning must be based on feedback from the consequences of behaviour that can be evaluated in terms of a set of criteria. These criteria are provided by the goal function. Learned modification of the goal function could not, of course, be based on information about the cost function, unless we are prepared to allow that the animal understands the evolutionary function of its own behaviour. In other words, the goal function is a fixed property of the robot which cannot be modified by learning, except where the learning is entirely preprogrammed (McFarland and Houston, 1981). Preprogrammed learning, of the kind discussed above could involve a contingent change in the goal function. Thus a very sophisticated robot could embody a rule such that when in a French kitchen it automatically switched to a preordained goal function.

If learning which depends upon feedback about the consequences of behaviour cannot be based upon changes in the goal function, it must be based on those parameters (of the plant equations) which alter the availability and accessibility of resources (see McFarland and Houston, 1981, chapter 10). In some situations the progress that an animal is likely to make by such learning is predictable, and the animal effectively has a choice between learning to exploit a new situation, or remaining with a known and mastered situation. The animal may also have a choice in the extent to which learning should be directed at modifications in the availability or accessibility aspects of the consequences of its behaviour (Tovish, 1982). In situations in which the animal is not able to anticipate its future learning progress, its task is complicated by the need to gain information. The optimal solution to this type of problem generally involves sampling as well as exploitation of resources (see Stephens and Krebs, 1986).

In considering the robot (or animal) as an experimenter, we are concerned with associative learning, by means of which the robot assesses the consequences of its own behaviour. In the final analysis, and by means of some (unknown) intervening mechanism, the robot must assign more or less weight to the behaviour that produces better consequences. Thus if using a new mop produces results that are better than usual, more value will be assigned to that activity. This means that the behaviour will be more likely to occur in the future. Similarly, if washing the dishes more quickly produces more breakages, less value will be assigned to that aspect of behaviour.

This is merely Thorndike's (1913) Law of Effect, which is useful as an empirical summary, but says nothing about the mechanisms involved in learning. In exploring these mechanisms, Dickinson (1980) considers the nature of the internal representation that encodes the learning experience. He distinguishes between procedural representations and declarative representations of knowledge. A procedural representation directly reflects the use to which the knowledge will be put in controlling the animal's behaviour, whereas a declarative representation corresponds to a statement describing relationships among events in the animal's world which does not commit the animal to any particular course of action. The distinction corresponds to that of 'knowing how' and 'knowing that' (see also Chapter 2).

In the case of instrumental learning (the kind of learning we want our robot to achieve), Dickinson (1980) notes that 'instrumental responses appear to be purposeful being directed at the goal of achieving the environmental change . . . The obvious way to find out whether information about the goal of action is represented in the underlying associative structure is to look at an animal's integrative capacity' (p. 103). He notes that a large number of experiments have attempted to do just this, and discusses the example of an experiment by Adams (then unpublished). Dickinson concludes (p. 109) that 'Adams' experiment on the devaluation of rewards leaves us with little doubt that integration can occur and that some form of declarative representation is also required'. I have no quarrel with this conclusion, but he also notes (p. 105) that 'extended practice of an instrumental act seems to produce a transition of control from the declarative to the procedural form, thereby setting up a habit. A habit can be viewed as an action we perform automatically in a given situation without direct reference to the goal of that action.' If we accept that instrumental learning may involve declarative representations that enable the animal to integrate the goal and the relevant behaviour, does this necessarily mean that we have to accept (as Dickinson does) that the goal-representation controls the behaviour?

Returning to our robot, if we accept that an explicit, declarative goal-representation is required for satisfactory instrumental learning, do we have to accept that the robot's behaviour is goal-directed? Suppose that clean windows is a desirable goal. The robot has an explicit declarative representation of clean windows. That is, it knows what clean windows look like. The question is how to attain this state of affairs. For Dickinson, the robot learns that polishing produces clean windows, it wants clean windows, so it polishes in order to attain clean windows. In other words, its behaviour is directed by the clean-window representation. This means that the robot will perform the behaviour (polishing) necessary to clean the window, so long as the behaviour continues to produce the desired results. If the polishing cloth became contaminated with oil so that polishing no longer made the window cleaner, then the robot would presumably stop polishing and take some corrective action. This is fine so long as the robot is merely a window-cleaning machine, but problems arise (as discussed in Chapter 5) if it also has other activities in its repertoire.

An alternative scenario is that the robot reviews its behavioural options and sees, not which is most likely to attain clean windows, but which is more likely to reduce the overall cost, or provide most utility. To do this the robot would have to be able to evaluate each potential activity in relation to its consequences. The most simple way to do this is to sample each activity and assess the consequences of each. This is likely to be both costly and dangerous. Suppose, however, that the robot could assess the consequences cognitively, without actually performing each activity. In some (familiar) circumstances, the robot would know the consequences of each activity, while in others it would not know but might be able to assess the cost of discovering the consequences. Thus the robot would be in the situation of an animal (discussed above) with a choice between remaining with a mastered situation, and learning to exploit a new situation. In other words, the robot would be designed to survey the range of (expected) utilities offered by the possible courses of action, and choose that with the greatest utility.

To achieve this design, the robot would have to have a declarative representation of the consequences of each activity. In other words it would have to know that the consequences of polishing a window were such and such (in terms of probable time and energy expenditure, wear and tear, and final condition of the window), and assign a quantity of utility to these. I am not suggesting that this involves conscious processes. Indeed, we make such assessments unconsciously on an everyday basis. Consider, for example, lunching in a cafeteria. You are hungry, the food is expensive and there is a long queue. You queue-up, select your lunch and pay for it, in advance of eating anything. You will probably not select too little, because you are hungry, and because you do not want to queue a second time. You will probably not select too much, because the food is expensive and you cannot afford to waste any. You will select just the right amount to satisfy your hunger. To do this you have to be able to assess the hunger-satisfying potential of a great variety of foods. Intuitively, you have to assess how many sausages are equivalent to a piece of cake, how much bread is equivalent to a salad. Moreover, your hunger is not a simple scalar quantity, but a vector (carbohydrate, protein, fat etc), so you have to assess the utility of each item in many dimensions, and trade this off against the price of each item. Each item displayed on the cafeteria shelves is judged on the basis of its price and utility, then a selection is made. The judgement is in terms of the consequences, both for stomach and pocket, of consuming the selected items.

The point here is that the behaviour is controlled, not by the goal-representation, but by trade-off considerations. This conclusion need not imply that explicit (declarative) representations do not exist, but rather that they do not control behaviour, and therefore the behaviour is not goal-directed (see Chapter 5). Dickinson may (or may not) well be correct in concluding that rats must make use of declarative representations in integrating knowledge about goals and behaviour. In fact, I see no reason why there should not be quite complicated cognitive processes involved in both associative learning and in the control of behaviour. My point is that these cognitive states have a role in the control of behaviour that is no different from that of other (non-cognitive) motivational states. They simply provide new stimuli for the trade-off process.

An animal (or robot), that can assess the extent to which the consequences of its current, or future, behaviour are beneficial, is a cognitive animal, provided the assessment involves declarative knowledge of the consequences. (The main constraint on this process will be the degree of observability of the consequences of the behaviour). In the non-learning robot, described in Chapter 5, the assessment is purely procedural, because the robot simply follows a predetermined set of rules. In some types of animal learning, however, it appears (from the work of Dickinson, Mackintosh, and others) that explanation in purely procedural terms is not sufficient. It is, therefore, reasonable to expect that a learning robot will have to make use of declarative representations. This need not mean, however, that the robot's behaviour must be purposive or goal-directed. Indeed, if the role of a goal-representation is to control behaviour, it is involved in a well-defined procedure, and should be defined as a procedural representation. The alternative, I suggest, is that the robot has declarative knowledge of the consequences of its various possible activities, which is built up by experience, and which is made use of in assessing the behavioural options. The behaviour of such a robot may appear, to an outside observer, to be goal-directed, because it will often be goal-achieving (see Chapter 5). Much is in the eye of the beholder, and we must ask why we have such a predeliction to assume that goal-achieving behaviour is purposive.

Communication

Suppose we imagine two robots sharing the household chores. To do this effectively, they would have to communicate with each other. Let us suppose that they were designed to communicate in the English language. What would they have to say to each other?

Much would depend on whether the robots were designed to compete or co-operate. If one robot had full information of the internal state of the other, it could behave as though the other's state were part of its own state, and could organize its behaviour accordingly (assuming the two goal functions were the same). This would be the ideal situation for co-operating robots, but not for competing robots. As in animal behaviour, it is in the interests of competing robots to control the information about internal state that is made available to others. Honesty is not always the best policy. It is in the interest of the individual robot to save time and energy by allowing the other robot to carry the greater burden of the household chores. As in the case of the single robot, we can expect there to be a trade-off between the benefits of keeping the house clean (to prevent competitive invasion from another species of robot) and the benefits of economizing (thus costing less than a rival of the same species).

'What are you going to do next?' says robot A to robot B. An honest robot might reply 'My top priority, as determined by my internal state and my perception of the kitchen, is to do the washing up. My second priority is to make the beds, and my third priority is to clean the kitchen floor. However, because of the cost of going upstairs, I will probably do the washing up, and then clean the kitchen floor before making the beds.' A less honest robot might reply: 'I intend to do the washing up'. If this were a complete lie, then robot A would soon learn that B was not to be believed, and B's utterances would become devalued. If the washing up were indeed B's top priority, then it might be better to say 'I intend to do the washing up' than to be completely honest. If B did not complete the washing up, and was challenged by A, B could say that the statement of intention was genuine, but that B had subsequently realised that some other job should be done instead. In fact, B may have known all along that the washing up would not remain B's top priority for long.

The honest robot would be at a disadvantage in the robot community. It would be less efficient than its competitors, and would be discriminated against in the market-place. It would be less efficient because other, less honest, robots would manipulate it into taking bad decisions, in the apparent spirit of co-operation. The transparently dishonest robot would be discriminated against by other robots, who would soon learn (or their designers would learn) to recognize a cad in the community. They would be unable to co-operate with others, and so would be less competitive in the market-place. The somewhat dishonest robot would sell well in the market-place.

My point is that the teleological mode of communication is somewhat dishonest. It can summarize what the actor is likely to do, but leaves room for manoeuvre. Thus a robot that says 'I intend to clean the floor next' is not revealing its true state. The statement may be true at the time, but the robot knows that when the time comes it will have to refuel. Moreover, if cleaning the floor ceases to be the most advantageous task, the robot can easily make an excuse for not doing it. 'I intended to clean the floor, but. . .' Thus, I am suggesting that the honest robot is exploited, the dishonest robot is avoided, and the teleological robot is the most successful at co-operating with other robots.

The question of honesty in animal communication has received considerable attention (Trivers, 1985; de Waal, 1982, 1986; Mitchell and Thompson, 1986; Rohwer and Rohwer, 1978). It is clear that, from an evolutionary point of view, there are situations in which animals are designed to deceive others, and so gain sexual, social, or political advantage. Indeed, it has been suggested that self-deception, hiding the truth from the conscious mind the better to hide it from others, has evolved in humans to fulfill these functions (Trivers, 1985). This line of reasoning has a number of implications for our concepts of motivation and goal-directed behaviour in animals.

Firstly, it further undermines the anthropomorphic approach to research in to animal behaviour. It makes little sense to ask whether animals have intentions similar to ours, if there is no rational basis for believing that we do have intentions. In other words, if our belief that our behaviour is (sometimes) intentional, is the result of evolutionarily designed self-deception, it may make sense to ask whether animals have similar illusions of intention (or of causality, or other cognitive phenomena), but not whether their behaviour is intentional.

Secondly, it raises the question of the extent to which deceptive behaviour is ever truly cognitive, perhaps even in humans. This is an issue which few authors face up to, an exception being de Waal in an excellent study of chimpanzees at Arnhem Zoo (1982). De Waal suggests (1986) that the process of gaining confidence in cognitive explanations is based mainly on human intuition.

> In this chapter I describe examples of nonritualised, intelligent forms of deception among semi-captive chimpanzees. The incidents discussed suggest, but cannot prove, the existence of intentional deception, since many of these examples lend themselves both to complex cognitive and to simpler explanations. . . One factor which seems to have some effect on a scientist's attitude about the controversial issue of animal mentality is the amount of experience he or she has had with the behaviour of nonhuman primates, especially pongids. The fact that absolute "nonbelievers" are rare among people familiar with members of these species means that direct exposure to their actions has the power to convince. Rather than on the gathering of explicit evidence, this process of gaining confidence in cognitive explanations is based mainly on human intuition (p. 221).

Perhaps we are designed (by natural selection) to assume that deceitful acts by fellow humans are intentional. We can readily translate such assumptions to our observations of chimpanzee behaviour, less readily to invertebrate behaviour. 'The look in the eyes of a primate often seems to tell the difference between intended and unintended effects of its actions on others' (p. 222). A scientist's hunch is acceptable as a start, provided that it leads to a theory that can be rejected in the face of the evidence. This has not yet been achieved in the field of animal cognition.

In ordinary parlance, we may deceive others fortuitously, or we may deceive others intentionally. We now have a third possibility, namely that we frequently deceive others as part of our normal (evolutionarily designed) behaviour, but sometimes we construe this as being intentional. This is a point I discuss further elsewhere (McFarland, 1988).

Finally, whatever the role of the teleological mode of communication in humans, and it may differ from individual to individual, I question whether any causal behavioural role is played by any explicit representation of the intended goal. The behaviour of others seems to us to be intentional, because we communicate in teleological terms, partly as a shorthand and partly as a cover for our true motives. Our own behaviour seems to us to be intentional, because we are designed to think in teleological terms. This mode of thinking is useful in interpreting the behaviour of our political rivals, but it is inherently self-deceiving.

Bibliography

Adam K 1980 Sleep as a restorative process and a theory to explain why. *Progress in Brain Research* 53: 289–306

Adam K, Oswald I 1977 Sleep is for tissue restoration. *Journal Royal College of Physicians* 11: 376–88

Adams C D, Dickinson A 1981a Instrumental responding following reinforcer devaluation. *Quarterly Journal Experimental Psychology* 33B: 109–12

Adams C D, Dickinson A 1981b Actions and habits: variations in associative representation during instrumental learning. In Spear N E, Miller R R (eds) *Information processing in animals: memory mechanisms*. Erlbaum, Hillsdale, N J

Adolph E F 1972 Some general concepts of physiological adaptations. In Yousef M K, Horvath S M, Bullard R W *Physiological adaptations* (eds). Academic Press, New York

Allison J 1979 Demand economics and experimental psychology. *Behavioral Science* 24: 403–15

Allison J 1983 *Behavioral economics*. Praeger Publishing, New York

Allison T, Cicchetti D V 1976 Sleep in mammals: ecological and constitutional correlates. *Science* 194: 732–4

Allison T, van Twyver H 1970 The evolution of sleep. *Natural History* 79: 56–65

Amlaner C J Jr, Ball N J 1983 A synthesis of sleep in wild birds. *Behaviour* 87: 85–119

Amlaner C J Jr, Ball N J 1988 Avian sleep. In Kryger M N (ed) *Principles and practice of sleep medicine*

Amlaner C J Jr, McFarland D J 1981 Sleep in the herring gull (*Larus argentatus*). *Animal Behaviour* 29: 551–6

Anderson F S 1968 Sleep in moths. *Opuscula Entomologica* 33: 15–24

Andrew R J 1956 Normal and irrelevant toilet behaviour in *Emberiza* spp. *British Journal Behaviour* 4: 85–91

Arbib M A 1966 Automata theory and control theory – a rapprochement. *Automatica* 3: 161–89

Arbib M A 1969 Automata theory. In Kalman R E, Falb P L, Arbib M A *Topics in mathematical system theory*. McGraw-Hill, New York, pp 163–233

Aschoff J 1960 Exogenous and endogenous components in circadian rhythms. *Cold Spring Harbor Symp. Quantitative Biology* 25: 11–28

Ball N J, Shaffery J P, Amlaner C J Jr 1984 Sleeping gulls and predator avoidance. *Animal Behaviour* 32: 1253–66

Barnard C J 1984 *Producers and scroungers*. Chapman and Hall, London

Barnard C J, Stephens H 1981 Prey size selection in lapwings in lapwing/gull associations. *Behaviour* 77: 1–22

Bateson P P G 1978 Early experience and sexual preferences. In Hutchison J B (ed) *Biological determinants of sexual behaviour*. John Wiley, London, pp 29–53

Blurton-Jones N G, Sibly R M 1978 Testing adaptiveness of culturally determined behaviour: do Bushman women maximise their reproductive success by spacing births widely and foraging seldom? In Reynolds V, Blurton-Jones N (eds) *Human behaviour and adaptation*. Taylor and Francis, London

Boakes R 1984 *From Darwin to behaviourism*. Cambridge University Press, Cambridge

Bolles R C 1970 Species-specific defense reaction and avoidance learning. *Psychological Review* 77: 32–48

Bolles R C 1975 *Theory of motivation* 2nd edn. Harper and Row, New York

Braithwaite R B 1946 Teleological explanation. *Proc. Aristotelian Society* XLVII: 1946–7

Braithwaite R B 1953 *Scientific explanation*. CUP, Cambridge

Brambell F W R 1965 (chairman) Report of the Technical Committee to enquire into the *Welfare of animals kept under intensive livestock husbandry systems*. Cmnd 2836. HMSO, London

Bunn D 1982 *Analysis for optimal decisions*. John Wiley, Chichester

Byrne R W, Whiten A 1985 Tactical deception of familiar individuals in baboons (*Papio ursinus*). *Animal Behaviour* 33: 669–73

Byrne R W, Whiten A 1988 *Machiavellian intelligence: social expertise and the evolution of intellect in monkeys, apes and humans*. Oxford University Press, Oxford

Caldwell R L 1986 The deceptive use of reputation by stomatopods. In Mitchell R W, Thompson N S (eds) *Deception*. SUNY Press, New York

Clutton-Brock T H, Guinness F E, Albon S D 1982 *Red deer: behaviour and ecology of two sexes*. University of Chicago Press, Chicago

Clutton-Brock T H, Harvey P H 1984 Comparative approaches to investigating adaptation. In Krebs J H, Davies N (eds) *Behavioural Ecology*. Blackwell Scientific Publications, Oxford

Corning W J, Dyal J A, Lahue R 1976 Intelligence: an invertebrate perspective. In Masterton R B, Hodos W, Jerison H (eds) *Evolution, brain and behaviour: persistent problems*. Lawrence Erlbaum Associates, Hillsdale, NJ, pp 215–63

Czeisler C A, Zimmerman J C, Ronda J M, Moore-Ede M C, Weitzman E D 1980 Timing of REM sleep is coupled to the circadian rhythm of body temperature in man. *Sleep* 2: 329–46

Daan S 1981 Adaptive daily strategies in behaviour. In Ashoff J (ed) *Handbook of behavioural neurobiology* vol. 4 *Biological rhythms*. Plenum Press, New York

Daan S 1982 Recuperative sleep as an adaptive behaviour. In Koella W P (ed) *Sleep*. Plenum Press, New York, 1–3

Darwin C 1872 *The expression of the emotions in man and the animals*. John Murray, London

Dawkins C R 1986 *The blind watchmaker*. Longman, Harlow

Dawkins M S 1980 *Animal suffering*. Chapman and Hall, London

Dawkins M S 1983 Battery hens name their price: consumer demand theory and the measurement of ethological 'needs'. *Animal Behaviour* 31: 1195–205

Dawkins R 1976 *The selfish gene*. Oxford University Press, Oxford

Dawkins R 1980 Good strategy or evolutionarily stable strategy? In Barlow G W, Silverberg J (eds) *Sociobiology: beyond nature/nurture?* Westview Press, Boulder, pp 331–67

Dawkins R, Krebs J R 1978 Animal signals: information or manipulation? In Krebs J R, Davies N B (eds) *Behavioural ecology*. Blackwell Scientific Publications, Oxford

Dawkins R, Krebs J R 1979 Arms races between and within species. *Proc. Royal Society London B* **205**: 489–511

de Waal F 1982 *Chimpanzee politics: power and sex among apes*. Harper and Row, New York

de Waal F 1986 Deception in the natural communication of chimpanzees. In Mitchell R W, Thompson N S (eds). *Deception: Perspectives on human and non human deceit*. State University of New York Press, Albany NY

Dennett D C 1983 Intentional systems in cognitive ethology: the 'Panglossian paradigm' defended. *Behavioral and Brain Sciences* **6**: 343–90

Dennett D C 1987 *The intentional stance*. MIT Press/A Bradford Book, Cambridge, Massachusetts

Dickinson A 1980 *Contemporary animal learning theory*. Cambridge University Press, Cambridge

Dickinson A 1985 Actions and habits: the development of behavioural autonomy. *Philosophical Transactions Royal Society London B* **308**: 67–78

Elgar M A, Harvey P H 1987 Basic metabolic rates in mammals: allometry, phylogeny and ecology. *Functional Ecology* **1**: 25–36

Elgar M A, Pagel M D, Harvey P H 1988 Sleep in mammals. *Animal Behaviour* (in press)

Emlen S T 1984 Cooperative breeding in birds and mammals. In Krebs J R, Davies N B (eds) *Behavioural ecology*. Blackwell Scientific Publications, Oxford, pp 305–39

Fishbein W 1981 *Sleep, dreams and memory*. MTP Press, Lancaster

Fodor J 1987 *Psychosemantics*. MIT Press/A Bradford Book, Cambridge, Massachusetts

Foulkes D 1983 Cognitive processes during sleep: evolutionary processes. In Mayes A (ed) *Sleep mechanisms and functions*. Van Nostrand Reinhold, Wokingham

Frese M, Sabini J (eds) 1985 *Goal directed behaviour: the concept of action in psychology*. Lawrence Erlbaum Associates, Hillsdale, NJ

Gould J L 1980 Sun compensation by bees. *Science* **207**: 545–7

Gould J L 1981 Language. In McFarland D J (ed) *The Oxford companion to animal behaviour*. Oxford University Press, Oxford

Griffin D R 1976, 1981 *The question of animal awareness*. The Rockefeller University Press, New York

Halliday T R, Slater P J B (eds) 1983 *Animal behaviour* 1. *Causes and effects*. Blackwell Scientific Publications, Oxford

Hamilton W D, Zuk M 1982 Heritable true fitness and bright birds: a role for parasites? *Science* **218**: 384–7

Harris M 1985 *Culture, people, nature* 4th edn. Harper & Row, New York

Hartmann E 1968 The 90-minute sleep–dream cycle. *Archives of General Psychiatry* **18**: 280–6

Heiligenberg W, Kramer U 1972 Aggressiveness as a function of external stimulation. *Journal Comparative Physiology* **77**: 332

Heller R, Milinski M 1979 Optimal foraging of sticklebacks on swarming prey. *Animal Behaviour* **27**: 1127–41

Hinde R A 1985 Was 'The expression of the emotions' a misleading phrase? *Animal Behaviour* **33**: 985–92

Hodos W 1982 Some perspectives on the evolution of intelligence and the brain. In Griffin D R (ed) *Animal mind – human mind*. Springer-Verlag, Berlin, Heidelberg, New York

Hogan J A, Kleist S, Hutchins C S L 1970 Display and food as reinforcers in the Siamese fighting fish (*Betta splendens*). *Journal Comparative Physiology Psychology* **70**: 351–7

Holland P C 1977 Conditioned stimulus as a determinant of the form of the Pavlovian conditioned response. *Journal Experimental Psychology: Animal Behaviour Processes* **3**: 77–104

Holland P C, Straub J J 1979 Differential effects of two ways of devaluing the unconditioned stimulus after Pavlovian appetitive conditioning. *Journal Experimental Psychology: Animal Behaviour Processes* **5**: 65–78

Hollard V D, Delius J D 1983 Rotational invariance in visual pattern recognition by pigeons and humans. *Science* **218**: 804–6

Honig W K 1978 On the conceptual nature of cognitive terms: an initial essay. In Hulse S H, Fowler H, Honig W K (eds) *Cognitive processes in animal behaviour*. Lawrence Erlbaum Associates, Hillsdale, NJ

Horne J A 1977 Factors relating to energy conservation during sleep in mammals. *Physiology Psychology* **5**: 403–8

Horne J A 1978 A review of the biological effects of total sleep deprivation in man. *Biological Psychology* **7**: 55–102

Horne J A 1983 Mammalian sleep function with particular reference to man. In Mayes A (ed) *Sleep mechanisms and functions*. Van Nostrand Reinhold, Wokingham

Horne J A 1987 *Why we sleep: perspectives on the function of sleep in humans and other mammals*. Oxford University Press, Oxford

Houston A I, McFarland D J 1980 Behavioural resilience and its relation to demand functions. In Staddon J E R (ed) *Limits to action: the allocation of individual behaviour*. Academic Press: New York, pp 177–203

Houston A I, McNamara J M 1988 A framework for the functional analysis of behavior. *Behavioural and Brain Sciences* (in press)

Houston A I, Sumida B 1985 A positive feedback model for switching between two activities. *Animal Behaviour* **33**: 315–25

Houston A I, Halliday T R, McFarland D J 1977 Towards a model of the courtship of the smooth newt *Triturus vulgaris*, with special emphasis on problems of observability in the simulation of behaviour. *Medical and Biological Engineering and Computation* **15**: 49–61

Howard I P 1982 *Human visual orientation*. Wiley, New York

Hughes B O, Duncan I J H 1981 Do animals have behavioural needs? *Applied Animal Ethology* **7**: 381–93

Humphrey N K 1978 Nature's psychologists. *New Scientist* **78**: 900

Humphrey N 1979 Nature's psychologists. In Josephson B, Ramachandra B S (eds) *Consciousness and the physical world*. Pergamon, New York. Reprinted in Sunderland E, Smith M T 1980. *The exercise of intelligence*. Garland STPM Press, New York

Huntingford F 1984 *The study of animal behaviour*. Chapman and Hall, London, New York

Jouvet M 1978 Does a genetic programming of the brain occur during paradoxical sleep? In Buser P A, Rougel-Buser A (eds) *Cerebral correlates of conscious experience* Elsevier Science Publishers, Amsterdam

Jouvet M 1980 Paradoxical sleep and the nature – nurture controversy. *Progress in Brain Research* 53: 331–46

Kacelnic A 1979 The foraging efficiency of great tits (*Parus major*) in relation to light intensity. *Animal Behaviour* 27: 237–41

Kalman R E 1963 Mathematical description of linear dynamical systems. *J.S.I.A.M. Control, Series A* 1: 152–92

Kalman R E 1968 New developments in systems theory relevant to biology. In Mesarovic M C (ed) *Systems theory and biology*. Springer-Verlag, Berlin

Krebs J R, McCleery R 1984 Optimisation in behavioural ecology. In Krebs J R, Davies N (eds) *Behavioural ecology. An evolutionary approach* 2nd edn. Blackwells Scientific Publications, Oxford

Larkin S 1981 Time and energy in decision-making. Unpublished D. Phil. thesis. University of Oxford

Larkin S, McFarland D J 1978 The cost of changing from one activity to another. *Animal Behaviour* 26: 1237–46

Le Magnen J 1985 *Hunger*. Cambridge University Press, Cambridge

Lea S E G 1978 The psychology and economics of demand. *Psychology Bulletin* 85: 441–66

Lea S E G, Tarpy R M 1986 Hamsters' demand for food to eat and hoard as a function of deprivation and cost. *Animal Behaviour* 34: 1759–68

Lea S E G, Roper T J 1977 Demand for food on fixed-ratio schedules as a function of the quality of concurrently available reinforcement. *Journal Experiment Analysis Behaviour* 27: 371–80

Lea S E G, Tarpy R M, Webley P 1987 *The individual in the economy*. Cambridge University Press, Cambridge

Lee R B 1972 The !Kung Bushmen of Botswana. In Bicchieri M G (ed) *Hunters and gatherers today*. Holt Rinehart Winston: New York

Lee R B 1979 *The !Kung San*. Cambridge University Press, Cambridge

Lee R B, DeVore I 1968 (eds) *Man the Hunter*. Aldine, Chicago

Lendrem D W 1983 Sleeping and vigilance in birds. I. Field observation of the mallard (*Anas platyrhynchos*). *Animal Behaviour* 31: 532–8

Lendrem D W 1984 Sleeping and vigilance in birds. II. An experimental study of the Barbary dove (*Streptopelia risoria*). *Animal Behaviour* 32: 243–18

Ludlow A R 1980 The evolution and simulation of a decision maker. In Toates F M, Halliday T R (eds) *Analysis of motivational processes*. Academic Press, London

Mackintosh N J 1983 *Conditioning and associative learning*. Clarendon Press, Oxford

Mackintosh N J, Dickinson A 1979 Instrumental (type II) conditioning. In Dickinson A, Boakes R A (eds) *Mechanisms of learning and motivation*. Erlbaum, Hillsdale, NJ

Maynard-Smith J 1978 Optimisation theory in evolution. *Annual Review Ecological Systems* 9: 31–56

Maynard-Smith J 1982 *Evolution and the theory of games*. Cambridge

University Press, Cambridge

McCleery R 1977 On satiation curves. *Animal Behaviour* 25: 1005–15

McCleery R 1978 Optimal behaviour sequences and decision making. In Krebs J R, Davies N B *Behavioural ecology*. Blackwells Scientific Publications, Oxford

McFarland D J 1965 The effect of hunger on thirst-motivated behaviour in the Barbary dove. *Animal Behaviour* 13: 292–300

McFarland D J 1970a Recent developments in the study of feeding and drinking in animals. *Journal Psychosomatic Research* 14: 229–37

McFarland D J 1970b Adjunctive behaviour in feeding and drinking situations. *Revue Comportement des Animals* 4: 64–73

McFarland D J 1971 *Feedback mechanisms in animal behaviour*. Academic Press, London

McFarland D J 1974 Time-sharing as a behavioral phenomenon. In Lehrman D S, Rosemblatt J S, Hinde R A, Shaw E (eds) *Advances in The Study of Behaviour*. Academic Press, New York

McFarland D J 1977 Decision-making in animals. *Nature, London* 269: 15–21

McFarland D J 1978 Hunger in interaction with other aspects of motivation. In Booth D A (ed) *Hunger models: computable theory of feeding control*. Academic Press, London, pp 375–405

McFarland D J 1981, 1987 ed *The Oxford companion to animal behaviour*. Oxford University Press, Oxford

McFarland D J 1983 Functional analysis of competing homeostatic systems. In Hales J R S (ed) *Thermal physiology*. Raven Press, New York

McFarland D J 1985 *Animal behaviour*. Pitman Books, London

McFarland D J 1988 Goals, no goals and own goals. In Montefiore A, Noble D (eds) *Goals, no goals and own goals*. Hutchinsons, London, pp. 83–8

McFarland D J, Houston A I 1981 *Quantitative ethology: the state space approach*. Pitman Books, London

McFarland D J, McFarland F J 1968 Dynamic analysis of an avian drinking response. *Medical and Biological Engineering* 6: 659–68

McFarland D J, Sibly R M 1972 'Unitary drives' revisited. *Animal Behaviour* 20: 548–63

McFarland D J, Sibly R M 1975 The behavioural final common path. *Philosophical Transactions Royal Society (Series B)* 270: 265–93

McGrath M J, Cohen D B 1978 REM sleep facilitation of adaptive waking behaviour: a review of the literature. *Psychology Bulletin* 85: 24–57

McNamara J M, Mace R H, Houston A I 1987 Optimal daily routines of singing and foraging in a bird singing to attract a mate. *Behavioural Ecology and Sociobiology* 20: 399–405

Meddis R 1975 On the function of sleep. *Animal Behaviour* 23: 676–91

Meddis R 1977 *The sleep instinct*. Routledge & Kegan Paul, London

Meddis R 1983 The evolution of sleep. In Mayes A (ed) *Sleep mechanisms and functions*. Van Nostrand Reinhold, Wokingham

Menzel E W, Wyers E J 1981 Cognitive aspects of foraging behaviour. In Kamil A C, Sargent T D (eds) *Foraging behaviour – ecological, ethological and psychological approaches*. Garland STPM Press, New York

Milinski M, Heller R 1978 Influence of a predator on the optimal foraging behaviour of sticklebacks (*Gasterosteus aculeatus* L). *Nature, London* 275: 642–4

Milsum J H 1966 *Biological control systems analysis.* McGraw-Hill, New York

Mitchell R W, Thompson N S 1986 *Deception.* SUNY, Albany, New York

Mogensen G J, Calaresu F R 1978 Food intake considered from the viewpoint of systems analysis. In Booth D (ed) *Hunger models.* Academic Press, London and New York.

Moore E F 1956 Gedanken-experiments on sequential machines. In Shannon C E, McCarthy J (eds) *Automata Studies.* Princeton University Press, Princeton NJ, pp 129–53

Morgan C L 1894 *Introduction to comparative psychology.* Scott, London

Mukhametov L M, Polyakova I G 1981 EEG investigation of sleep in porpoises (*Phocoena phocoena*). *Journal Higher Nervous Activity SSR* (Russian) **31**: 333–9

Nagel E 1961 *The structure of science.* Routledge & Kegan Paul, London

Nagel T 1974 What is it like to be a bat? *Philosophical Reviews* **83**: 435–50

Niebuhr V 1981 An investigation of courtship feeding in herring gulls, *Larus argentatus*. *Ibis* **123**: 218–23

Nisbet I C T 1977 Courtship feeding and clutch size in common terns, *Sterna hirundo*. In Stonehouse B, Perrins C M (eds) *Evolutionary ecology.* Macmillan, London

Noble D 1988 Intentional action and physiology. In Montefiore A, Noble D (eds) *Goals, no goals and own goals.* Hutchinsons, London

Opp M R, Ball N J 1985 Bioenergetic consequences of sleep based on time budgets of free-ranging glaucous-winged gulls. *Sleep Research* **14**: 21

Opp M R, Ball N J, Miller D E, Amlaner C J 1987 Thermoregulation of sleep: effects of thermal stress on sleep patterns of glaucous-winged gulls (*Larus glaucescens*). *Journal of Thermal Biology* **12**: 199–202

Pagel M D, Harvey P H 1988 The comparative method. *Quarterly Reviews of Biology* (in press)

Parmeggiani P L 1977 Interaction between sleep and thermoregulation. *Waking and Sleeping* **1**: 123–32

Perry G C 1978 Behavioural needs. *The Veterinary Record* **102**: 386–7

Petrusic W A, Varro L, Jamieson D G 1978 Mental rotation validation of two spatial ability tests. *Psychological Research* **40**: 139–48

Rachlin H 1980 Economics and behavioral psychology. In Staddon J E R (ed) *Limits to action.* Academic Press, New York

Rechtschaffen A, Gilliland M A, Bergmann B M, Winter J B 1983 Physiological correlates of prolonged sleep deprivation in rats. *Science* **221**: 182–4

Rohwer S, Ewald P W 1981 The cost of dominance and advantage of coordination in a badge signalling system. *Evolution* **35**: 441–54

Rohwer S, Rohwer F C 1978 Status signalling in Harris sparrows: experimental deception achieved. *Animal Behaviour* **26**: 1012–22

Rolls B J, Rolls E T 1982 *Thirst.* Cambridge University Press, Cambridge

Romanes G J 1882 *Animal intelligence.* Kegan Paul, Trench, London

Rosen R 1967 *Optimality principles in biology.* Butterworths, London

Rosenblueth A, Wiener W, Bigelow J 1943 Behavior, purpose and teleology. *Philosophy of Science* **10**: 18–24

Rowell C H F 1961 Displacement grooming in the chaffinch. *Animal Behaviour* **9**: 38–63

Rozin P 1976a The evolution of intelligence and access to the cognitive unconscious. *Progress in Psychobiology and Physiological Psychology* **6**: 245–76

Rozin P 1976b The selection of food by rats, humans and other animals. In Rosenblatt J S, Hinde R A, Shaw E, Beer C (eds) *Advances in the study of behaviour* vol. 6. Academic Press, New York, pp 21–76

Rozin P, Kalat J 1971 Specific hungers and poison avoidance as adaptive specialisations of learning. *Psychological Reviews* 78: 459–86

Rusak B 1981 Vertebrate behavioral rhythms. In Aschoff J (ed) *Handbook of behavioural neurobiology* vol. 4, *Biological rhythms*. Plenum Press, New York

Russell B 1921 *The analysis of mind.* Macmillan, New York

Scudder T 1971 Gathering among African woodland savannah cultivators; a case study: The Gwembe Tonga. *Zambian Papers*, 5: University of Zambia, Institute of African Studies

Sevenster P 1961 A causal analysis of a displacement activity (fanning in *Gasterosteus aculeatus* L.). *Behaviour* Supplement IX

Shaffery J P, Ball N J, Amlaner C J Jr 1985 Manipulating daytime sleep in herring gulls (*Larus argentatus*). *Animal Behaviour* 33: 566–72

Sherry D F, Mrosovsky N, Hogan J A 1980 Weight loss and anorexia during incubation in birds. *Journal Comparative Physiology Psychology* 441: 89–98

Sibly R M 1984 Models of producer/scrounger relationships between and within species. In Barnard C J (ed) *Producers and scroungers.* Croom Helm, London and Sydney; Chapman and Hall, New York

Sibly R M, McCleery R H 1983 Increase in weight of herring gulls while feeding. *Journal Animal Ecology* 52: 35–50; The distribution between feeding sites of herring gulls breeding at Walney Island, UK. 52: 51–68

Sibly R M, McCleery R H 1985 Optimal decision rules for gulls. *Animal Behaviour* 33: 449–65

Sibly R M, McFarland D J 1974 A state-space approach to motivation. In McFarland D J (ed) *Motivational control systems analysis.* Academic Press, London and New York, pp 213–50

Sibly R M, McFarland D J 1976 On the fitness of behaviour sequences. *American Naturalist* 110: 601–17

Spellerberg I F, Hoffmann K 1972 Circadian rhythm in lizard critical minimum temperature. *Naturwissenschaften* 59: 517–18

Stephens D W, Krebs J R 1986 *Foraging theory.* Princeton University Press, Princeton, NJ

Thompson D B A, Barnard C J 1983 Anti-predator responses in mixed species flocks of lapwings, golden plovers and black-headed gulls. *Animal Behaviour* 31: 585–93

Thorndike E L 1913 *The psychology of learning. (Educational psychology II).* Teachers College, New York

Thorpe W H 1965 The assessment of pain and distress in animals. In Brambell F W R (chairman) *Report of the technical committee to enquire into the welfare of animals kept under intensive livestock systems.* Cmnd 2836. HMSO, London

Tinbergen N 1951 *The study of instinct.* Clarendon Press, Oxford

Toates F M 1980 *Animal behaviour – a systems approach.* John Wiley, Chichester

Toates F M 1986 *Motivational systems.* Cambridge University Press, Cambridge

Toutain P, Ruckebusch Y 1972 Secretions nasolabiales au cours du sommeil paradoxal chez les bouins. *C. R. Acad. Sc.*, Paris. **274**: 2519–22

Tovish A 1982 Learning to improve the availability and accessibility of resources. In McFarland D (ed) *Functional ontogeny.* Pitman Books, London

Trivers R L 1971 The evolution of reciprocal altruism. *Quarterly Review Biology* **46**: 35–57

Trivers R L 1972 Parental investment and sexual selection. In Campbell B (ed) *Sexual selection and the descent of man.* Aldine, Chicago

Trivers R L 1985 *Social evolution.* Benjamin/Cummings Publishing, Menlo Park, California

Tuddenham R D 1963 The nature and measurement of intelligence. In Postman L (ed) *Psychology in the making: histories and selected research problems.* Knopf, New York, pp 469–525

Vestergaard K 1980 The regulation of dust-bathing and other patterns in the laying hen: a Lorenzian approach. In: Moss R (ed) *The laying hen and its environment.* Martinus Nijhoff, The Hague, pp 101–20

Vestergaard K 1982 Dust-bathing in the domestic fowl – diurnal rhythm and dust deprivation. *Applied Animal Ethology* **8**: 487–95

von Cranach M, Harre R 1982 *The analysis of action.* Cambridge University Press, Cambridge

von Ruppell G 1986 A "lie" as a directed message of the arctic fox (*Alopex lagopus* L.) In Mitchell R W, Thompson N S (eds) *Deception.* SUNY Press, New York

Walker J M, Berger R J 1980 Sleep as an adaptation for energy conservation functionally and physiologically related to hibernation and shallow torpor. *Progress Brain Research* **53**: 255–78

Walker J M, Glotzbach S F, Berger R J, Heller H C 1977 Sleep and hibernation in ground squirrels (*Citellus* spp): electrophysiological observations. *American Journal Physiology* **233**: R213–21

Webb W B 1975 *Sleep, the gentle tyrant.* Prentice-Hall, Englewood Cliffs

Webb W B 1983 Theories in modern sleep research. In Mayes A (ed) *Sleep mechanisms and functions.* Van Nostrand Reinhold, Wokingham

Whiten A, Byrne R W 1988 Tactical deception in primates. *Behavioural and Brain Sciences* (in press)

Wiepkema P 1968 Behaviour changes in CBA mice as a result of one goldthioglucose injection. *Behaviour* **32**: 179–210

Wilkinson G S 1984 Reciprocal food sharing in the vampire bat. *Nature* **308**: 181–4

Wilson D M 1966 Insect walking. *Annual Review Entomology* **11**: 103–22

Wirtshafter D, Davis J D 1977 Set points, settling points, and the control of bodyweight. *Physiology and Behaviour* **19**, 75–78

Woodfield A 1976 *Teleology.* Cambridge University Press, Cambridge

Wooton R J 1976 *The biology of sticklebacks.* Academic Press, London

Ydenberg R C, Houston A I 1986 Optimal trade-offs between competing behavioural demands in the great tit. *Animal Behaviour* **34**: 1041–50

Zepelin H, Rechtschaffen A 1970 Mammalian sleep and longevity. *Psychophysiology* **7**: 306

Zepelin H, Rechtschaffen A 1974 Mammalian sleep, longevity and energy metabolism. *Brain Behaviour and Evolution* **10**: 425–70

Index